Association for Women in Mathematics Series

Volume 20

Series Editor
Kristin Lauter
Microsoft Research
Redmond,Washington, USA

T0172128

Association for Women in Mathematics Series

Focusing on the groundbreaking work of women in mathematics past, present, and future, Springer's Association for Women in Mathematics Series presents the latest research and proceedings of conferences worldwide organized by the Association for Women in Mathematics (AWM). All works are peer-reviewed to meet the highest standards of scientific literature, while presenting topics at the cutting edge of pure and applied mathematics, as well as in the areas of mathematical education and history. Since its inception in 1971, The Association for Women in Mathematics has been a non-profit organization designed to help encourage women and girls to study and pursue active careers in mathematics and the mathematical sciences and to promote equal opportunity and equal treatment of women and girls in the mathematical sciences. Currently, the organization represents more than 3000 members and 200 institutions constituting a broad spectrum of the mathematical community in the United States and around the world.

More information about this series at http://www.springer.com/series/13764

Carolina Araujo • Georgia Benkart
Cheryl E. Praeger • Betül Tanbay
Editors

World Women in Mathematics 2018

Proceedings of the First World Meeting
for Women in Mathematics (WM)^2

WORLD MEETING
FOR WOMEN
IN MATHEMATICS

Editors

Carolina Araujo
Instituto Nacional de Matemática Pura e
Aplicada (IMPA)
Rio de Janeiro, Brazil

Georgia Benkart
Department of Mathematics
University of Wisconsin–Madison
Madison, WI, USA

Cheryl E. Praeger
Department of Mathematics
University of Western Australia
Crawley, WA, Australia

Betül Tanbay
Department of Mathematics
Boğaziçi University
Bebek, Istanbul, Turkey

ISSN 2364-5733   ISSN 2364-5741   (electronic)
Association for Women in Mathematics Series
ISBN 978-3-030-21172-1   ISBN 978-3-030-21170-7   (eBook)
https://doi.org/10.1007/978-3-030-21170-7

Mathematics Subject Classification: 00B20

This Springer imprint is published by the registered company Springer Nature Switzerland AG.
The registered company address is: Gewerbestrasse 11, 6330 Cham, Switzerland

Dedicated to the memory of Maryam Mirzakhani, a shining light of inspiration for women in mathematics.

Preface

The International Congress of Mathematicians (ICM), the most important meeting of the international mathematical community, is coordinated every four years by the International Mathematical Union (IMU). ICM 2018 took place in Rio de Janeiro, Brazil, from August 1 to August 9, 2018. The first World Meeting for Women in Mathematics - $(WM)^2$ - was held in Rio de Janeiro on July 31, 2018, as a satellite event of ICM 2018. Conceived by the IMU Committee for Women in Mathematics (CWM), with a focus on Latin America, the World Meeting for Women in Mathematics brought together over 300 mathematicians from 51 nations to celebrate women mathematicians and also to reflect on gender issues in mathematics, challenges, initiatives, and perspectives for the future. The meeting was complemented by the panel discussion organized by CWM on August 2, as part of the ICM 2018 program. This volume, organized in coordination with the Association for Women in Mathematics (AWM), records the first World Meeting for Women in Mathematics and the CWM panel discussion in ICM 2018.

The first part of the volume is devoted to the World Meeting for Women in Mathematics. It starts with a short report on the activities of the $(WM)^2$, including pictures that attest to the lively and friendly atmosphere of the meeting. Following the report, survey research papers from four of the invited lecturers provide a panoramic view of different fields in pure and applied mathematics. The first paper, by Etienne de Klerk and Monique Laurent, is anchored in the keynote lecture delivered by Laurent. It is a thorough survey on the generalized problem of moments, a class of linear conic infinite dimensional optimization problems that arise in many areas of applied mathematics. In the next article, Alicia Dickenstein addresses biochemical reaction networks and how techniques from computational and real algebraic geometry have been successfully applied in recent years to analyze them. The paper by Stella Brassesco and Maria Eulália Vares builds on the lecture given by Vares. It gives an introduction to the stochastic modeling of metastability, a very frequent phenomenon in nature, which also finds many applications in science and engineering. Part 1 then closes with the inviting note by Maria J. Esteban, based on her public lecture at the World Meeting for Women in Mathematics, entitled "How mathematics is changing the world."

The second part of the volume documents the CWM panel discussion at ICM 2018. The panel, "The gender gap in mathematical and natural sciences from a historical perspective," was chaired by Caroline Series, and featured contributions from Marie-Francoise Roy (chair of CWM), June Barrow-Green, and Silvina Ponce-Dawson. The paper by Helena Mihaljević and Marie-Françoise Roy traces the footprints of women lecturers in the International Congress of Mathematicians since its inception. In particular, it pictures the first two women to give plenary lectures at an ICM, Emmy Noether in ICM 1932 in Zurich and Karen Uhlenbeck in ICM 1990 in Kyoto. (While this book was being prepared, we received the welcome news that Karen Uhlenbeck was awarded the 2019 Abel Prize!) June Barrow-Green's essay investigates the historical context of the gender gap in mathematics, analyzing challenges faced by women mathematicians during the last two hundred and fifty years, and shedding light on some of the problems still encountered today. This part closes with the paper by Silvina Ponce Dawson, which describes a series of actions taken by the International Union of Pure and Applied Physics to reduce the gender gap and increase diversity and inclusion in physics.

The organization of the $(WM)^2$ was greatly inspired by the shining light of Maryam Mirzakhani, the first woman mathematician to be awarded the Fields Medal, at ICM 2014 in Seoul. As a tribute, the CWM created *Remember Maryam Mirzakhani*, a memorial exhibition of 18 original posters portraying the late mathematician. Inaugurated at the $(WM)^2$, the exhibition remained open during ICM 2018. Since then, it has been shown in several venues. Maryam Mirzakhani will always be a beacon for women in mathematics, and we dedicate this volume to her memory.

Rio de Janeiro, Brazil   Carolina Araujo
March 2019

Acknowledgments

The World Meeting for Women in Mathematics was part of the Biennium of Mathematics in Brazil, officially sponsored by the Brazilian Development Bank (BNDES). It received financial/logistical support from the Committee for Women in Mathematics (CWM) of the International Mathematical Union (IMU), Institute for Pure and Applied Mathematics (IMPA), Brazilian Mathematical Society (SBM), Brazilian National Council for Scientific and Technological Development (CNPq), São Paulo Research Foundation (FAPESP), Serrapilheira, and L'Oréal Brasil. We thank these institutions for their support.

We thank the members of the organizing committee of the World Meeting for Women in Mathematics for the dedicated work to make the event happen and the members of the program committee for the excellent job in selecting the speakers for the meeting. We thank the invited lecturers for their efforts in making their presentations accessible to the general audience and the panelists for sharing their experience as members of active networks of women mathematicians. We thank the organizing committee of the ICM 2018 for the constant support and partnership and the staff of IMPA for the invaluable assistance with the organization. We thank Thaís Jordão and Rafael Meireles Barroso for their work on *Remember Maryam Mirzakhani* memorial exhibition. We thank the Simons Foundation for sponsoring the film "Journeys of Women in Mathematics" and the Association for Women in Mathematics (AWM) for the collaboration in creating this volume. We thank Aydın Tibet for the $(WM)^2$ logo and Bel Junqueira for the photography coverage.

Finally, we thank all the participants of the World Meeting for Women in Mathematics for their enthusiasm, making July 31, 2018, a day to remember.

Contents

List of Contributors

Carolina Araujo Instituto Nacional de Matemática Pura e, Aplicada (IMPA), Rio de Janeiro, Brazil

June Barrow-Green The Open University, Milton Keynes, UK

Stella Brassesco Instituto Venezolano de Investigaciones Científicas, Caracas, Venezuela

Silvina Ponce Dawson Departamento de Física, FCEN-UBA, Ciudad Universitaria, Buenos Aires, Argentina
IFIBA, CONICET-UBA, Ciudad Universitaria, Buenos Aires, Argentina

Etienne de Klerk Tilburg University, Tilburg, The Netherlands
Delft University of Technology, Delft, The Netherlands

Alicia Dickenstein Department of Mathematics, FCEN, University of Buenos Aires, Buenos Aires, Argentina
IMAS, UBA-CONICET, University of Buenos Aires, Buenos Aires, Argentina

Maria J. Esteban CEREMADE (CNRS UMR n° 7534), Université Paris-Dauphine, PSL Research University, Paris 16, France

Monique Laurent CWI Amsterdam, Amsterdam, The Netherlands
Tilburg University, Tilburg, The Netherlands

Helena Mihaljević Hochschule für Technik und Wiertschaft Berlin, University of Applied Science, Wilhelminenhofstraße 75A, 12459 Berlin, Germany

Marie-Françoise Roy IRMAR (UMR CNRS 6625), Université de Rennes 1, Rennes Cedex, France

Maria Eulália Vares Instituto de Matemática, Universidade Federal do Rio de Janeiro, Rio de Janeiro, RJ, Brazil

Part I
The First World Meeting for Women in Mathematics

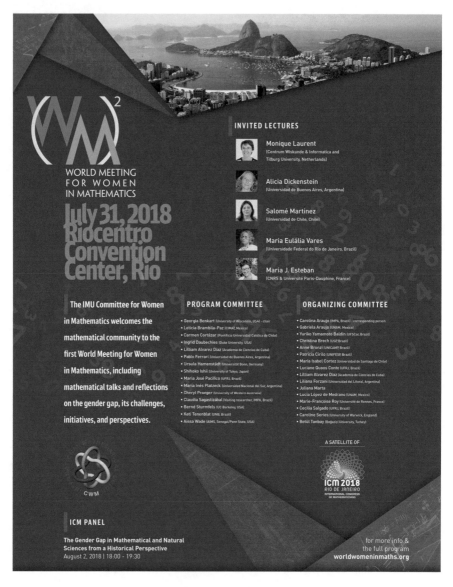

Fig. P1   Poster of the World Meeting for Women in Mathematics

Report on Activities of the First World Meeting for Women in Mathematics

Carolina Araujo

Abstract This is a short report of activities of the World Meeting for Women in Mathematics, which took place in Rio de Janeiro, Brazil, on July 31, 2018, as a satellite event of the ICM 2018.

1 Attendance

The first World Meeting for Women in Mathematics - $(WM)^2$ - took place in Rio de Janeiro, Brazil, on July 31, 2018, as a satellite event of the ICM 2018. It had a total of 296 registered participants, mostly women, plus around 50 guest participants (accompanying people and participants of ICM 2018) (Fig. 1). Registered participants came from 51 different countries, distributed as follows.

Fig. 1 Participants of the $(WM)^2$

C. Araujo (✉)
Instituto Nacional de Matemática Pura e Aplicada (IMPA), Rio de Janeiro, RJ, Brazil
e-mail: caraujo@impa.br

© The Association for Women in Mathematics and the Author(s) 2019
C. Araujo et al. (eds.), *World Women in Mathematics 2018*, Association for Women in Mathematics Series 20, https://doi.org/10.1007/978-3-030-21170-7

Algeria—6	Indonesia—8	Russia—1
Argentina—21	Iran—7	Senegal—2
Australia—2	Japan—2	Serbia—4
Belarus—1	Kyrgyzstan—1	Slovenia—1
Brazil—123	Mexico—5	South Africa—1
Burkina Faso—1	Moldova—2	South Korea—2
Cambodia—1	Montenegro—2	Spain—2
Cameroon—1	Morocco—1	Tanzania—1
Canada—3	Mozambique—1	Tunisia—2
Chile—8	Nepal—2	Turkey—5
Colombia—2	Netherlands—1	Ukraine—5
Congo—1	Nigeria—8	UK—1
Denmark—1	Pakistan—2	USA—11
Ecuador—1	Peru—2	Uruguay—1
France—11	Philippines—5	Uzbekistan—1
Ghana—2	Poland—1	Venezuela—1
India—17	Romania—1	Vietnam—1

2 Program at a Glance

7:00–9:00	Registration
9:00–9:20	World Premiere of the film "Journeys of Women in Mathematics"
9:25–10:10	Keynote Lecture: Monique Laurent "Convergence Analysis of Approximation Hierarchies for Polynomial Optimization"
10:15–10:30	Memorial for Maryam Mirzakhani
10:35–11:20	Lecture: Alicia Dickenstein "Algebra and Geometry in the Study of Enzymatic Cascades"
11:25–12:25	Group discussions
12:30–14:20	Lunch + Posters
14:25–15:10	Lecture: Salomé Martínez "Reaction-Diffusion Equations, Population and Gender Dynamics"
15:15–16:00	Lecture: Maria Eulália Vares "Revisiting the Contact Process"
16:00–16:40	Coffee break
16:40–17:25	Public Lecture: Maria J. Esteban "Why Mathematics Is Changing the World"
17:30–18:45	Panel discussion "Networks of Women in Mathematics"
19:00	ICM opening cocktail reception

Fig. 2 Left: The protagonists of the first part of the film "Journeys of Women in Mathematics." Right: Poster of the full length film

3 Journeys of Women in Mathematics

The program of $(WM)^2$ opened with the world premiere of the film "Journeys of Women in Mathematics." The film was created by the IMU Committee for Women in Mathematics, filmed and edited by Micro-Documentaries, and made possible by a grant from the Simons Foundation. The first part of the film features Carolina Araujo from Brazil, Neela Nataraj from India, and Aminatou Pecha from Cameroon (Fig. 2). It describes their research, the mathematical aspirations, successes and barriers faced by women in their region, all told in the words of the women themselves.

The second part of the film was shot during $(WM)^2$ and ICM 2018. It gives a lively presentation of the atmosphere at the meeting, and features interviews of six women in mathematics from Latin America: Alicia Dickenstein from Argentina, Natalia García-Colín from Mexico, Salomé Martínez from Chile, Jaqueline Godoy Mesquita from Brazil, Carolina Neira Jiménez from Colombia, and Maria Eulália Vares from Brazil. The final full-length version of the film is available at the following webpage.

https://www.mathunion.org/fileadmin/CWM/Videos/Simons_Foundation-
WM2_Conference-1363-02-T04_FINAL_Stitch.mp4.

4 Scientific Lectures

The scientific program of $(WM)^2$ included five 45-minute plenary lectures: keynote lecture by Monique Laurent (Netherlands), invited lectures by Alicia Dickenstein (Argentina), Salomé Martínez (Chile) and Maria Eulália Vares (Brazil), and public lecture by Maria J. Esteban (France). The lectures were aimed at a general audience of mathematicians, providing panoramic views of different fields in pure and applied mathematics.

1. Monique Laurent "Convergence Analysis of Approximation Hierarchies for Polynomial Optimization"

   Abstract: We consider the polynomial optimization problem, which asks to minimize a multivariate polynomial f over a compact semi-algebraic set K. Equivalently, this is asking to find a measure with positive density function, which minimizes the expected value of f over K. This is a hard problem, which has spurred a booming research activity in the past two decades, starting with seminal works by Lasserre and Parrilo in 2000 and onward. In a nutshell, results from real algebraic geometry about positive polynomials and from functional analysis about moments of measures are used to construct hierarchies of bounds that converge to the global minimum of f over K. These bounds are based on using sums-of-squares positivity certificates. While testing positivity of a polynomial is a hard computational problem, the key fact is that there exist efficient algorithms to search for sums of squares of polynomials.

   In this lecture we will focus on hierarchies of upper bounds, that are obtained by selecting sums-of-squares density functions with growing degrees d. We will discuss several recent results about the convergence rate of these hierarchies. For general convex bodies K we can show a convergence rate in $O(1/d)$ and, for simpler sets like the hypercube, we can show a stronger convergence rate in $O(1/d^2)$. In addition this convergence analysis is tight, which relies on establishing links to orthogonal polynomials and their extremal roots.

   This lecture is based on joint work with Etienne de Klerk.

2. Alicia Dickenstein "Algebra and Geometry in the Study of Enzymatic Cascades"

   Abstract: In recent years, techniques from computational and real algebraic geometry have been successfully used to address mathematical challenges in systems biology. The algebraic theory of chemical reaction systems aims to understand their dynamic behavior by taking advantage of the inherent algebraic structure in the kinetic equations, and does not need a priori determination of the parameters, which can be theoretically or practically impossible.

   I will give a gentle introduction to general results based on the network structure. In particular, I will describe a general framework for biological systems, called MESSI systems, that describe Modifications of type Enzyme-Substrate or Swap with Intermediates, and include many post-translational modification networks. I will also outline recent methods to address the important question of multistationarity, in particular in the study of enzymatic cascades, and will point out some of the mathematical challenges that arise from this application.

3. Salomé Martínez "Reaction-Diffusion Equations, Population and Gender Dynamics"

   Abstract: Reaction-diffusion models have been widely used to study fundamental questions in population dynamics. This type of partial differential equation provides a way to translate local assumptions regarding the movement, growth and interactions of the individuals of a species, into global features of the population giving us a theoretical framework for questions such as the persistence of a species, invasions, coexistence of competing populations. Different mathematical tools from nonlinear analysis and dynamical systems can be used to study the consequences of varying different population characteristics have in the long term dynamics.

In this talk we will study competitive reaction-diffusion systems of the form:

$$\begin{cases} \dfrac{\partial u}{\partial t} = Lu + u(m(x) - u - bv) \ \text{ in } \Omega, \ t > 0, \\[2ex] \dfrac{\partial u}{\partial t} = Mv + v(m(x) - cu - v) \ \text{ in } \Omega, \ t > 0, \\[2ex] \nabla \dfrac{u}{m} \cdot \hat{n} = \nabla v \cdot \hat{n} = 0 \ \text{ on } \partial\Omega, \ t > 0, \end{cases}$$

with u, v representing the densities of two competing populations in an isolated habitat Ω, $a(x)$ the space dependent per-capita growth rate, $b, c > 0$ accounting for competition coefficients, and L and M elliptic operators accounting for the dispersal strategies of each species. In particular, we will discuss how the relationship between population dispersal and competition affects the persistence, dispersal and coexistence of the species.

In this talk we will also explore some issues related to the persistence and dispersal of women in STEM in an environment where they account for less than 17% of the population. I will share how we have been able to significantly grow and thrive through the formation and strengthening of networks and alliances. In particular, we will discuss the process that led to the creation of the Direction for Diversity and Gender, the first in a Faculty of Sciences, Math, and Engineering in Chile, which I currently lead.

4. Maria Eulália Vares "Revisiting the Contact Process"

Abstract: Introduced by T. Harris more than forty years ago, the classical contact process is a simple stochastic model to describe the propagation of an infection in a population, where the individuals sit on the vertices of a graph, also called sites. It can be thought as a Markov process on the space of subsets of the set of all sites, identifying the state "infected" or "healthy" of each individual. Its description is simple and in the most natural examples the model shows interesting features, like dynamical phase transition and metastability, which have been precisely described. This process can be described through paths in a random space-time graph, also called Harris system. Several variations have been considered recently, including the case where the sites are given by the vertices of a random graph, or the contact process with two types of individuals. In this talk I would like to describe another variation, where one loses the Markov property but for which the investigation of phase transition, thought in terms of percolation properties, remains interesting. This is based on joint work with L.R. Fontes, D. Marchetti, and T. Mountford. If time allows I would like to discuss features of the metastable behavior of a contact process with two types of individuals, and which is work done by my PhD student at UFRJ, Mariela P. Machado.

5. Maria J. Esteban "Why Mathematics Is Changing the World"
   Abstract: Mathematics has always been key to help understanding the world in
   which we live. But it is becoming more and more one of the key technologies
   to change and improve it, and to foster innovation. Mathematics is behind most
   of the important recent technological developments. This is due to the increasing
   complexity of the processes that need to be described and understood. The use
   of sophisticated mathematics together with advanced algorithms lead to efficient
   and robust methods to solve problems that would be out or reach otherwise. This
   lecture will be devoted to the presentation of examples showing the strength of
   the mathematical technologies behind this immense success.

5   Group Discussions

About 200 participants divided up into 15 small discussion groups. Each group
focused on one topic, and used a common language of their choice (one of Arabic,
English, French, Portuguese or Spanish). The discussions lasted for about one
hour, and were moderated by one or two people per group, who had previously
volunteered as facilitators. At the end of the discussion, each group was invited to
formulate a proposition or a question. The conclusions were presented during the
panel discussion at the end of the program (Fig. 3).

Fig. 3 Group discussions at the $(WM)^2$

The topics and the group facilitators (all languages included) were:

- Diversity in Mathematics (Eloah Oliveira Corrêa, Maria Isabel Cortez, Manjusha Majumdar, Fatma Zohra Nouristrategies, Gabriela Araujo Pardo, Eliane Costa Santos, Valdirene Rosa de Souza);
- Gender gap in mathematics (Zamurat Ayobami Adegboye);
- Strategies to encourage women to do research in mathematics (Karina Batistelli, Eunice Mureithi, Neela Nataraj, Socorro Rangel, Selmane Schehrazad, Luz De Teresa);
- Maternity and career (Adeniji Adenike);
- Mentoring early career (Mary Durojaye, Mercedes Siles Molina);
- Patriarchal practices in Academia (Fadipe-Joseph Olubunmi);
- Public policies to promote women in science (Yuliya Mishura, Fagueye Ndiaye, Mythily Ramaswamy, Catherine Roberts);
- Strategies to stimulate undergraduate girls in mathematics (Mercy Gyamea Amankwah, Cristina Lizana Araneda, Yuriko Baldin, Ogunrinde Roseline Bosede, Jyoti U. Devkota, Walcy Santos);
- Female role models and how to highlight women's contributions (Aruquia Peixoto);
- The status of female researchers in the math community (Atinuke Adebanji, Stefanella Boatto).

The propositions and questions elaborated by the discussion groups have been synthesized by the organizers of the $(WM)^2$ into the following four topics.

1. *Strategies to stimulate girls to pursue undergraduate studies in mathematics.*

   The importance of starting to promote the gender balance in early stages of education, and the key role played by teachers in this process were highlighted. Proposed strategies include: to offer courses on popularization/applications of mathematics in secondary school; to fight against the preconceived idea that mathematics is a subject not suited for girls; to encourage girls to study mathematical sciences; to organize summer camps in mathematics either entirely for girls or with a good gender balance; to stimulate, maintain and expand scientific initiation programs for undergraduate and high school students. Teachers should encourage and value the participation of women studying mathematics; be aware of the cultural barriers that impair girls' performance in mathematics, and help them break those; diffuse the role of women as protagonists throughout the history of mathematics; organize regular seminars where women's contributions are discussed; and stimulate girls to be more vocal. It has been pointed out that gender balance could be achieved with complementary and not competitive interactions. The following question was posed: What is the ideal ratio of boys to girls in programs designed to motivate girls in mathematics?

2. *Strategies to encourage women to do research in mathematics, to highlight contributions by women, and to establish role models.*
   Three main aspects were discussed:

   a. The need to increase the perception/visibility of women's research work in mathematics. Proposed strategies include: to cite the work of women during classes and lectures; to encourage women to write more text books, to participate in events, to propose projects and to participate in scientific societies; to have more local seminars so women who have travel restrictions can be up to date in their research fields; to provide more travel and research grants, scholarships and childcare at scientific events.
   b. The need to promote collaboration and to value work between women. The importance of having mentoring networks for women mathematicians (with senior mathematicians mentoring young researchers, young researchers mentoring Ph.D students, and so on) was highlighted.
   c. The need to create and strengthen support networks for women mathematicians, as a way for women to develop courage and self-confidence. Role models play a role in these support networks, as does sharing experiences such as balancing family, career, research and responsibilities.

3. *Public policies to promote women in science and to overcome the gender gap in mathematics.*
   Different strategies were proposed to promote and retain women in science, including: to use positive discrimination, that is, give preference to a woman between two candidates at equal level competing for the same position; alternation between men and women in decision making positions; to include maternity in women's CV and take this into account when evaluating their applications, acknowledging the greater impact of maternity in women's career as compared to men's; to increase the number of nominations of females to scientific committees and prize awards. The question of whether women have to fit in the patriarchal model in order to survive in academia, with stereotyped roles and behaviours, was discussed. The importance of the CWM led project "A Global Approach to the Gender Gap in Mathematical and Natural Sciences", which aims at a better documentation of the gender gap in science, was highlighted.

4. *Strategies to encourage diversity.*
   It was observed that in mathematics (and academia more generally), linear trajectories in research are more valued, and usually better suited for men. The importance of recognizing and valuing the diversity of identities and trajectories, including those focused on aspects other than research, such as popularization of mathematics, organizing events, teaching, etc, was noted. Social and ethnic diversity in mathematics was also discussed, and the difficulty of experiencing diversity in mathematics in practice was acknowledged. Some positive examples were discussed, such as the ethno-mathematics studies in Mozambican culture, the laboratory of mathematical teaching at UFBA (Federal University of Bahia, Brazil) and the ethno-mathematics group at UNILAB and UFABC (Federal University of ABC, Brazil).

Fig. 4 Poster session at the $(WM)^2$

6 Poster Presentations

The poster session was a very active part of the program, with the presentation of 57 research posters covering various areas of mathematics, as well as 14 mathematic posters describing initiatives and statistics about women in mathematics worldwide. The complete list of posters presented at $(WM)^2$, including titles, authors and abstracts, can be found in the following webpages (Fig. 4).

Mathematical posters:

https://impa.br/wp-content/uploads/2019/03/WM2-PS_Matematica.pdf.

Thematic posters about women in mathematics:

https://impa.br/wp-content/uploads/2019/03/WM2-PS_Mulheresnamatematica.
pdf.

7 Panel Discussion

The program of $(WM)^2$ ended with a 75-minute panel discussion about "Networks of Women in Mathematics", moderated by Carolina Araujo. The panelists were:

- Christina Brech (Brazil)—Brazilian network
- Natalia Garcia (Mexico)—Latin American networks

Fig. 5 Panel discussion "Networks of Women in Mathematics"

- Magnhild Lien (USA)—Association for Women in Mathematics (AWM)
- Marie Françoise Ouedraogo (Burkina Faso)—African Women in Mathematics (AWMA)
- Marie-Françoise Roy (France)—European Women in Mathematics (EWM)
- Riddhi Shah (India)—Indian Women in Mathematics (IWM)

panelists shared briefly the history of the networks of women mathematicians in their regions, their challenges and most successful initiatives, making clear the importance of networks to improve the situation of women in mathematics (Fig. 5).

Following the presentations, facilitators from the morning group discussions presented their previously formulated propositions, and the program closed with attendees voting by a large majority to celebrate women in mathematics on May 12, starting in 2019.

8   Program Committee

- Georgia Benkart (University of Wisconsin, USA)—chair
- Leticia Brambila-Paz (CIMAT, Mexico)
- Carmen Cortázar (Pontificia Universidad Católica de Chile)
- Ingrid Daubechies (Duke University, USA)
- Lilliam Alvarez Diaz (Academia de Ciencias de Cuba)
- Pablo Ferrari (Universidad de Buenos Aires, Argentina)
- Ursula Hamenstädt (Universität Bonn, Germany)

- Shihoko Ishii (University of Tokyo, Japan)
- Maria José Pacifico (UFRJ, Brazil)
- María Inés Platzeck (Universidad Nacional del Sur, Argentina)
- Cheryl E. Praeger (University of Western Australia)
- Claudia Sagastizábal (Visiting researcher, IMPA, Brazil)
- Bernd Sturmfels (UC Berkeley, USA)
- Keti Tenenblat (UNB, Brazil)
- Aissa Wade (AIMS, Senegal/Penn State, USA)

9   Organizing Committee

- Carolina Araujo (IMPA, Brazil)—corresponding person
- Gabriela Araujo (UNAM, Mexico)
- Yuriko Yamamoto Baldin (UFSCar, Brazil)
- Christina Brech (USP, Brazil)
- Anne Bronzi (UNICAMP, Brazil)
- Patricia Cirilo (UNIFESP, Brazil)
- Maria Isabel Cortez (Universidad de Santiago de Chile, Chile)
- Luciane Quoos Conte (UFRJ, Brazil)
- Lilliam Alvarez Diaz (Academia de Ciencias de Cuba, Cuba)
- Liliana Forzani (Universidad del Litoral, Argentina)
- Juliana Marta
- Lucía López de Medrano (UNAM, Mexico)
- Marie-Françoise Roy (Université de Rennes, France)
- Cecilia Salgado (UFRJ, Brazil)
- Caroline Series (University of Warwick, England)
- Betül Tanbay (Boğaziçi University, Turkey)

10   A Tribute to Maryam Mirzakhani

The whole mathematical community was deeply saddened by the untimely death of Maryam Mirzakhani on July 14, 2017, at the age of forty. She was the first woman mathematician to be awarded the Fields Medal, at ICM 2014 in Seoul. As a tribute to her memory, CWM created *Remember Maryam Mirzakhani*, a memorial exhibition with 18 original posters portraying Maryam Mirzakhani, two volumes containing her mathematical work, one volume with articles about her, and a book of condolences for attendees to sign. The exhibition was signed by Thaís Jordão (curator) and Rafael Meireles Barroso (designer). The exhibition was inaugurated at the $(WM)^2$, and remained open during the ICM 2018. Since then, it has been shown in several venues (Fig. 6).

The Memorial for Maryam Mirzakhani during the $(WM)^2$ included a screening of the film shown in ICM 2014 when she received the Fields medal, followed by the following words by Betül Tanbay and one minute silence.

Fig. 6 *Remember Maryam Mirzakhani* memorial exhibition at the $(WM)^2$

Maryam Mizrakhani was born on May 12, 1977.
Her childhood was spent in Teheran during the Iran-Iraq war.
In an interview, Maryam said, "I was very lucky in many ways. The war ended when I finished elementary school; I couldn't have had the great opportunities that I had if I had been born ten years earlier."
Maryam went to Farzanegan, a high school for girls with exceptional talents. Her outstanding mathematical ability became evident when she won gold medals in the International Mathematical Olympiads in Hong Kong (1994) and in Canada (1995), achieving a perfect score.
In 1995, Maryam joined the Sharif University of Technology in Teheran, where she completed her Bachelor of Science (BS) in 1999. By then she had already published three papers, of which two were in graph theory.
Maryam worked for her PhD at Harvard under Fields medallist Curtis McMullen on hyperbolic surfaces. In 2004, she was awarded the Leonard M and Eleanor B Blumenthal award for her thesis, which was judged as outstanding and brought her a Clay fellowship at Princeton. Rising rapidly from assistant professor to professor, in 2008 she moved with her husband Jan Vondrák, a computer scientist, to a chair at Stanford.
Her daughter Anahita was born in 2011. Maryam has often been pictured with large sheets of paper spread on the floor on which she would visualise her mathematics of curved surfaces. Her daughter would remark that her mother was painting.
Maryam was an invited speaker at ICM 2010 in Hyderabad. She received innumerable awards and prizes for her mathematical work, and was awarded the Fields Medal at ICM 2014 in Seoul.

In the words of Terence Tao:

> *Her greatest recent achievement has been her "magic wand" theorem with Alex Eskin, which is basically the analogue of the famous measure classification and orbit closure theorems of Marina Ratner, in the context of moduli spaces instead of unipotent flows on homogeneous spaces. Ratner's theorems are fundamentally important to any problem to which a homogeneous dynamical system can be associated, as it gives a good description of the equidistribution of any orbit of that system; and it seems the Eskin-Mirzakhani result will play a similar role in problems associated instead to moduli spaces. The remarkable proof of this result uses almost all of the latest techniques that had been developed for*

Fig. 7 Left: Dr Ashraf Daneshkhah making the proposition to celebrate women in mathematics on May 12 (Marcos Arcoverde/ICM2018). Right: Voting at the closing of the meeting

homogeneous dynamics, and ingeniously adapts them to the more difficult setting of moduli spaces, in a manner that had not been dreamed of being possible only a few years earlier.[1]

An application of these results is to the "illumination problem." Imagine a room with mirrored walls. If a candle is placed at some location in the room, will it illuminate every other point in the room?
I have the feeling Maryam will be the candle illuminating every point in the space of mathematicians.

In the words of Caroline Series:

> *With her infectious enthusiasm, she was always keen to discuss mathematics, always optimistic about what could be done, modest and unassuming while projecting an unwavering self-confidence. She had a reputation for tackling the most difficult questions with dogged persistence.[2]*

Allow me to add as a woman mathematician from her neighbourhood that Maryam showed forever that excellence in mathematics is not a matter of gender nor geography. Mathematics being a universal value, it belongs to us all.
Her region did not see much peace.
May she rest in peace.

On behalf of the Women's Committee at the Iranian Mathematical Society, Dr Ashraf Daneshkhah then presented to the participants a proposal that Maryam Mirzakhani's birthday - May 12 - be celebrated worldwide within the mathematical community as "Women in Mathematics Day." The program of the $(WM)^2$ closed with attendees voting by a large majority to celebrate women in mathematics on May 12, starting in 2019 (Fig. 7).

Following this decision, there is currently an international initiative to celebrate women in mathematics on May 12, coordinated by several organizations for women in mathematics worldwide: African Women in Mathematics Association (AWMA), Association for Women in Mathematics (AWM), Colectivo de Mujeres

---

[1] Terence Tao's blog https://terrytao.wordpress.com.

[2] Series, Caroline. *Maryam Mirzakhani and her work.* Math. Today (Southend-on-Sea) 53 (2017), no. 5, 192–194.

Matemáticas de Chile, European Women in Mathematics (EWM), Indian Women and Mathematics (IWM), and Women's Committee of the Iranian Mathematical Society (Fig. 8).

https://may12.womeninmaths.org.

Fig. 8 "Door". Piece of *Remember Maryam Mirzakhani* memorial exhibition

A Survey of Semidefinite Programming Approaches to the Generalized Problem of Moments and Their Error Analysis

Etienne de Klerk and Monique Laurent

Abstract The generalized problem of moments is a conic linear optimization problem over the convex cone of positive Borel measures with given support. It has a large variety of applications, including global optimization of polynomials and rational functions, option pricing in finance, constructing quadrature schemes for numerical integration, and distributionally robust optimization. A usual solution approach, due to J.B. Lasserre, is to approximate the convex cone of positive Borel measures by finite dimensional outer and inner conic approximations. We will review some results on these approximations, with a special focus on the convergence rate of the hierarchies of upper and lower bounds for the general problem of moments that are obtained from these inner and outer approximations.

1 Introduction

The classical problem of moments is to decide when a measure is determined by a set of specified moments and variants of this problem were studied (in the univariate case) by leading nineteenth and early twentieth century mathematicians, like Hamburger, Stieltjes, Chebyshev, Hausdorff, and Markov. We refer to [1] for an early reference and to the recent monograph [51] for a comprehensive treatment of the moment problem.

The generalized problem of moments is to optimize a linear function over the set of finite, positive Borel measures that satisfy certain moment-type conditions.

E. de Klerk
Tilburg University, Tilburg, The Netherlands

Delft University of Technology, Delft, The Netherlands
e-mail: E.deKlerk@uvt.nl

M. Laurent (✉)
CWI Amsterdam, Amsterdam, The Netherlands

Tilburg University, Tilburg, The Netherlands
e-mail: M.Laurent@cwi.nl

© The Association for Women in Mathematics and the Author(s) 2019
C. Araujo et al. (eds.), *World Women in Mathematics 2018*, Association for Women in Mathematics Series 20, https://doi.org/10.1007/978-3-030-21170-7_1

More precisely, we consider continuous functions f_0 and f_i ($i \in [m]$) where $[m] = \{1, \ldots, m\}$, that are defined on a compact set $K \subset \mathbb{R}^n$. The generalized problem of moments (GPM) may now be defined as follows.[1]

Generalized Problem of Moments (GPM)

$$val := \inf_{\mu \in \mathcal{M}(K)_+} \left\{ \int_K f_0(x)d\mu(x) \ : \ \int_K f_i(x)d\mu(x) = b_i \quad \forall i \in [m] \right\}, \qquad (1)$$

where

- $\mathcal{M}(K)_+$ denotes the convex cone of positive, finite, Borel measures (i.e., Radon measures) supported on the set K[2];
- The scalars $b_i \in \mathbb{R}$ ($i \in [m]$) are given.

In this survey we will mostly consider the case where all f_i's are polynomials, and will always assume $K \subseteq \mathbb{R}^n$ to be compact. Moreover, for some of the results, we will also assume that K is a basic semi-algebraic set and we will sometimes further restrict to simple sets like a hypercube, simplex or sphere.

The generalized problem of moments has a rich history; see, e.g., [1, 30, 51] and references therein and [36] for a recent overview of many of its applications. In the recent years modern optimization approaches have been investigated in depth, in particular, by Lasserre (see [32], the monograph [33] and further references therein). Among others, there is a well-understood duality theory, and hierarchies of inner and outer approximations for the cone $\mathcal{M}(K)_+$ have been introduced that lead to converging upper and lower bounds for the problem (1). In this survey we will present these hierarchies and show how the corresponding bounds can be computed using semidefinite programming. Since several overviews are already available on general properties of these hierarchies (e.g., in [33, 34, 37, 38]), our main focus here will be on recent results that describe their rate of convergence. We will review in particular in more detail recent results on the upper bounds arising from the inner approximations, and highlight some recent links made with orthogonal polynomials and cubature rules for integration.

---

[1] We only deal with the GPM in a restricted setting; more general versions of the problem are studied in, e.g., [54].

[2] Formally, we consider the usual Borel σ-algebra, say $\mathcal{B}$, on $\mathbb{R}^n$, i.e., the smallest (or coarsest) σ-algebra that contains the open sets in $\mathbb{R}^n$. A positive, finite Borel measure μ is a nonnegative-valued set function on $\mathcal{B}$, that is countably additive for disjoint sets in $\mathcal{B}$. The support of μ is the set, denoted Supp(μ), and defined as the smallest closed set S such that $\mu(\mathbb{R}^n \setminus S) = 0$.

1.1 The Dual Problem of the GPM

The GPM is an infinite-dimensional conic linear program, and therefore it has an associated dual problem. Formally we introduce a duality (or pairing) between the following two vector spaces:

1. the space $\mathcal{M}(K)$ of all signed, finite, Borel measures supported on K,
2. the space $C(K)$ of continuous functions on K, endowed with the supremum norm $\|\cdot\|_\infty$.

The duality (pairing) in question is provided by the nondegenerate bilinear form $\langle\cdot,\cdot\rangle : C(K) \times \mathcal{M}(K) \to \mathbb{R}$, defined by

$$\langle f, \mu\rangle = \int_K f(x)d\mu(x) \quad (f \in C(K),\ \mu \in \mathcal{M}(K)).$$

Thus the dual cone of $\mathcal{M}(K)_+$ w.r.t. this duality is the cone of continuous functions that are nonnegative on K, and will be denoted by $C(K)_+ = (\mathcal{M}(K)_+)^*$.

In our setting of compact $K \subset \mathbb{R}^n$, $\mathcal{M}(K)$ is also the dual space of $C(K)$, i.e., $\mathcal{M}(K)$ may be associated with the space of linear functionals defined on $C(K)$. In particular, due to the Riesz-Markov-Kakutani representation theorem (e.g. [56, §1.10]), every linear functional on $C(K)$ may be expressed as

$$f \mapsto \langle f, \mu\rangle \quad \text{for a suitable } \mu \in \mathcal{M}(K).$$

As a result, we have the weak* topology on $\mathcal{M}(K)$ where the open sets are finite intersections of elementary sets of the form

$$\{\mu \in \mathcal{M}(K) \mid \alpha < \langle f, \mu\rangle < \beta\},$$

for given $\alpha, \beta \in \mathbb{R}$, and $f \in C(K)$, and the unions of such finite intersections.

A sequence $\{\mu_k\} \subset \mathcal{M}(K)$ converges in the weak* topology, say $\mu_k \rightharpoonup \mu$, if, and only if,

$$\lim_{k\to\infty} \langle f, \mu_k\rangle = \langle f, \mu\rangle \ \forall f \in C(K). \tag{2}$$

As a consequence of (2), the cone $\mathcal{M}(K)_+$ is closed and the set of probability measures in $\mathcal{M}(K)$ is closed.

By Alaoglu's theorem, e.g. [2, Theorem III(2.9)], the following set (i.e., the unit ball in $\mathcal{M}(K)$) is compact in the weak* topology of $\mathcal{M}(K)$:

$$\{\mu \in \mathcal{M}(K) \mid |\langle f, \mu\rangle| \le 1 \ \forall f \in C(K) \text{ with } \|f\|_\infty \le 1\}. \tag{3}$$

Hence the set of probability measures in $\mathcal{M}(K)$ is compact, since it is a closed subset of the compact set in (3), and thus it provides a compact base in the weak* topology for the cone $M(K)_+$. This implies again that $M(K)_+$ is closed in this topology (using Lemma 7.3 in [2, Part IV]) and we will also use this fact to analyze duality in the next section.

Dual Linear Optimization Problem of (1)

Using this duality setting, the dual conic linear program of (1) reads

$$val^* := \sup_{y \in \mathbb{R}^m} \left\{ \sum_{i \in [m]} b_i y_i \ : \ f_0 - \sum_{i \in [m]} y_i f_i \in C(K)_+ \right\},$$

$$= \sup_{y \in \mathbb{R}^m} \left\{ \sum_{i \in [m]} b_i y_i \ : \ f_0(x) - \sum_{i \in [m]} y_i f_i(x) \geq 0 \ \forall x \in K \right\}. \qquad (4)$$

By the duality theory of conic linear optimization, one has the following duality relations; see, e.g., [2, Section IV.7.2] or [33, Appendix C].

Theorem 1 *Consider the GPM (1) and its dual (4). Assume (1) has a feasible solution. One has val $\geq$ val* (weak duality), with equality val $=$ val* (strong duality) if the cone $\{(\langle f_0, \mu \rangle, \langle f_1, \mu \rangle, \ldots, \langle f_m, \mu \rangle) \ : \ \mu \in M(K)_+\}$ is a closed subset of $\mathbb{R}^{m+1}$. If, in addition, val $> -\infty$ then (1) has an optimal solution.*

We mention another sufficient condition for strong duality, that is a consequence of Theorem 1 in our setting.

Corollary 1 *Assume (1) has a feasible solution, and there exist $z_0, z_1, \ldots, z_m \in \mathbb{R}$ for which the function $\sum_{i=0}^m z_i f_i$ is strictly positive on K (i.e., $\sum_{i=0}^m z_i f_i(x) > 0$ for all $x \in K$). Then, val $=$ val* holds and (1) has an optimal solution.*

Hence, if in problem (1) we optimize over the probability measures (i.e., with $f_1 \equiv 1$, $b_1 = 1$) then the assumptions in Corollary 1 are satisfied.

We indicate how Corollary 1 can be derived from Theorem 1. Consider the linear map $L : M(K) \rightarrow \mathbb{R}^{m+1}$ defined by $L(\mu) = (\langle f_0, \mu \rangle, \ldots, \langle f_m, \mu \rangle)$, which is continuous w.r.t. the weak* topology on $M(K)$. First we claim Ker $L \cap M(K)_+ = \{0\}$. Indeed, assume $L(\mu) = 0$ for some $\mu \in M(K)_+$. Setting $f = \sum_{i=0}^m z_i f_i$, $L(\mu) = 0$ implies $\langle f, \mu \rangle = 0$ and thus $\mu = 0$ since f is strictly positive on K. Since the cone $M(K)_+$ has a compact convex base in the weak$*$ topology and the linear map L is continuous, we can conclude that the image $L(M(K)_+)$ is closed (using Lemma 7.3 in [2, Part IV]). Now we can conclude using Theorem 1.

1.2 Atomic Solution of the GPM

If the GPM has an optimal solution, then it has a finite atomic optimal solution, supported on at most m points (i.e., the weighted sum of at most m Dirac delta measures). This is a classical result in the theory of moments; see, e.g., [48] (univariate case), [29] (which shows an atomic measure with $m + 1$ atoms using induction on m) and a modern exposition in [54] (which shows an atomic measure with m atoms). The result may also be obtained as a consequence of the following, dimension-free version of the Carathéodory theorem.

Theorem 2 (See, e.g., Theorem 9.2 in Chapter III of [2]) *Let S be a convex subset of a vector space such that, for every line L, the intersection $S \cap L$ is a closed bounded interval. Then every extreme point of the intersection of S with m hyperplanes can be expressed as a convex combination of at most $m + 1$ extreme points of S.*

Atomic Solution of the (GPM)

Theorem 3 *If the GPM (1) has an optimal solution then it has one which is finite atomic with at most m atoms, i.e., of the form $\mu^* = \sum_{\ell=1}^{m} w_\ell \delta_{x^{(\ell)}}$ where $w_\ell \geq 0$, $x^{(\ell)} \in K$, and $\delta_{x^{(\ell)}}$ denotes the Dirac measure supported at $x^{(\ell)}$ ($\ell \in [m]$).*

This result can be derived from Theorem 2 in the following way. By assumption, the GPM has an optimal solution μ^*. Moreover, since it has one at an extreme point we may assume that μ^* is an extreme point of the feasibility region $\mathcal{M}(K)_+ \cap \cap_{i=1}^{m} H_i$ of the program (1), where H_i is the hyperplane $\langle f_i, \mu \rangle = b_i$. Then the following set $S = \{\mu \in \mathcal{M}(K)_+ : \mu(K) = \mu^*(K)\}$ meets the condition of Theorem 2, since the set of probability measures in $\mathcal{M}(K)_+$ is compact in the weak* topology, and any line in a topological vector space is closed (e.g. [2, p. 111]). Moreover, the extreme points of S are precisely the scaled Dirac measures supported by points in K (see, e.g., Section III.8 in [2]). In addition, μ^* is an extreme point of the set $S \cap \cap_{i=1}^{m} H_i$ and thus, by Theorem 2, μ^* is a conic combination of $m + 1$ Dirac measures supported at points $x^{(\ell)} \in K$ for $\ell \in [m + 1]$. Finally, as in [54], consider the LP

$$\min \sum_{\ell=1}^{m+1} w_\ell f_0(x^{(\ell)}) \text{ s.t. } w_\ell \geq 0 \ (\ell \in [m + 1]), \ \sum_{\ell=1}^{m+1} w_\ell f_i(x^{(\ell)}) = b_i \ (i \in [m])$$

whose optimal value is val. Then an optimal solution attained at an extreme point provides an optimal solution of the GPM (1) which is atomic with at most m atoms.

1.3 GPM in Terms of Moments

From now on we will assume the functions $f_0, f_1, \ldots, f_m$ in the definition of the
GPM (1) are all polynomials and the set K is compact. Then the GPM may be
reformulated in terms of the moments of the variable measure μ. To be precise,
given a multi-index $\alpha = (\alpha_1, \ldots, \alpha_n) \in \mathbb{N}^n$ the moment of order α of a measure
$\mu \in \mathcal{M}(K)_+$ is defined as

$$m_\alpha^\mu(K) := \int_K x^\alpha d\mu(x).$$

Here we set $x^\alpha = x_1^{\alpha_1} \cdots x_n^{\alpha_n}$. We may write the polynomials $f_0, f_1, \ldots, f_m$ in
terms of the standard monomial basis as:

$$f_i(x) = \sum_{\alpha \in \mathbb{N}_d^n} f_{i,\alpha} x^\alpha \quad \forall i = 0, \ldots, m,$$

where the $f_{i,\alpha} \in \mathbb{R}$ are the coefficients in the monomial basis, and we assume the
maximum total degree of the polynomials $f_0, f_1, \ldots, f_m$ to be at most d.

Throughout we let $\mathbb{N}_d^n = \{\alpha \in \mathbb{N}^n : |\alpha| \leq d\}$ denote the set of multi-indices,
with $|\alpha| = \sum_{i=1}^n \alpha_i$, and $\mathbb{R}[x]_d$ denotes the set of multivariate polynomials with
degree at most d.

GPM in Terms of Moments

We may now rewrite the GPM (1) in terms of moments:

$$\inf_{\mu \in \mathcal{M}(K)_+} \left\{ \sum_{\alpha \in \mathbb{N}_d^n} f_{0,\alpha} m_\alpha^\mu(K) : \sum_{\alpha \in \mathbb{N}_d^n} f_{i,\alpha} m_\alpha^\mu(K) = b_i \; \forall i \in [m] \right\}.$$

Here d is the maximum degree of the polynomials $f_0, f_1, \ldots, f_m$.

Thus we may consider the set of all possible truncated moments sequences:

$$\left\{ \left(m_\alpha^\mu(K) \right)_{\alpha \in \mathbb{N}_d^n} : \mu \in \mathcal{M}(K)_+ \right\},$$

and describe the inner and outer approximations for $\mathcal{M}(K)_+$ in terms of this set.

1.4 Inner and Outer Approximations

We will consider two types of approximations of the cone $M(K)_+$, namely inner and outer conic approximations.

Inner Approximations

The underlying idea, due to Lasserre [35], is to consider a subset of measures μ in $M(K)_+$ of the form

$$d\mu = h \cdot d\mu_0,$$

where h is a polynomial sum-of-squares density function, and $\mu_0 \in M(K)_+$ is a fixed reference measure with $\text{Supp}(\mu_0) = K$.

To obtain a finite dimensional subset of measures, we will limit the total degree of h to some value $2r$ where $r \in \mathbb{N}$ is fixed. The cone of sum-of-squares polynomials of total degree at most $2r$ will be denoted by Σ_r, hence

$$\Sigma_r = \left\{ \sum_{i=1}^{k} p_i^2 : k \in \mathbb{N},\, p_i \in \mathbb{R}[x]_r,\, i \in [k] \right\}.$$

In this way one obtains the cones

$$M_{\mu_0}^r := \{\mu \in M(K)_+ \ : \ d\mu = h \cdot d\mu_0,\ h \in \Sigma_r\} \quad (r = 1, 2, \ldots) \quad (5)$$

which provide a hierarchy of inner approximations for the set $M(K)_+$:

$$M_{\mu_0}^r \subseteq M_{\mu_0}^{r+1} \subseteq M(K)_+.$$

Outer Approximations

The dual GPM (4) involves the nonnegativity constraint

$$f_0(x) - \sum_{i=1}^{m} y_i f_i(x) \geq 0 \ \forall x \in K,$$

which one may relax to a sufficient condition that guarantees the nonnegativity of the polynomial $f_0 - \sum_{i=1}^{m} y_i f_i$ on K. Lasserre [31] suggested to use the following sufficient condition in the case when K is a basic closed semi-algebraic set, i.e., when we have a description of K as the intersection of the level sets of polynomials g_j ($j \in [k]$):

$$K = \left\{ x \in \mathbb{R}^n \ : \ g_j(x) \geq 0 \quad \forall j \in [k] \right\}.$$

Namely, consider the condition

$$f_0 - \sum_{i=1}^{m} y_i f_i = \sigma_0 + \sum_{j=1}^{k} \sigma_j g_j,$$

where each σ_j is a sum-of-squares polynomial and the degree of each term $\sigma_j g_j$ $(0 \leq j \leq k)$ is at most $2r$, so that the degree of the right-hand-side polynomial is at most $2r$. Here we set $g_0 \equiv 1$ for notational convenience. Thus we replace the cone $C(K)_+$ by a cone of the type:

$$Q^r(g_1, \ldots, g_k) := \left\{ f \; : \; f = \sigma_0 + \sum_{j=1}^{k} \sigma_j g_j, \; \sigma_j \in \Sigma_{r_j}, \; j = 0, 1, \ldots, k \right\},$$

(6)

where we set $r_j := r - \lceil \deg(g_j)/2 \rceil$ for all $j \in \{0, \ldots, k\}$.

The cone $Q^r(g_1, \ldots, g_k)$ is known as the *truncated quadratic module* generated by the polynomials $g_1, \ldots, g_k$. By definition, its dual cone consists of the signed measures μ supported on K such that $\int_K f d\mu \geq 0$ for all $f \in Q^r(g_1, \ldots, g_k)$:

$$(Q^r(g_1, \ldots, g_k))^* = \left\{ \mu \in \mathcal{M}(K) : \int_K f(x) d\mu(x) \geq 0 \quad \forall f \in Q^r(g_1, \ldots, g_k) \right\}.$$

(7)

This provides a hierarchy of outer approximations for the cone $\mathcal{M}(K)_+$:

$$\mathcal{M}(K)_+ \subseteq (Q^{r+1}(g_1, \ldots, g_k))^* \subseteq (Q^r(g_1, \ldots, g_k))^*.$$

We will also briefly consider the tighter outer approximations for the cone $\mathcal{M}(K)_+$ obtained by replacing the truncated quadratic module $Q^r(g_1, \ldots, g_k)$ by the larger cone $Q^r \left(\prod_{j \in J} g_j : J \subseteq [k] \right)$, thus the truncated quadratic module generated by all pairwise products of the g_j's (also known as the pre-ordering generated by the g_j's). Then we have

$$\mathcal{M}(K)_+ \subseteq \left(Q^r \left(\prod_{j \in J} g_j : J \subseteq [k] \right) \right)^* \subseteq (Q^r(g_1, \ldots, g_k))^*.$$

2   Examples of GPM

The GPM (1) has many applications. Below we will list some examples that are directly relevant to this survey; additional examples in control theory, option pricing in finance, and others, can be found in [32, 33, 36].

Global Minimization of Polynomials on Compact Sets

Consider the global optimization problem:

$$val = \min_{x \in K} p(x) \qquad (8)$$

where p is a polynomial and K a compact set. This corresponds to the GPM (1) with $m = 1$, $f_0 = p$, $f_1 = 1$ and $b_1 = 1$, i.e.:

$$val = \min_{\mu \in M(K)_+} \left\{ \int_K p(x) d\mu(x) \; : \; \int_K d\mu(x) = 1 \right\}.$$

In the following sections we will focus on deriving error bounds for this problem when using the inner and outer approximations of $M(K)_+$.

Global Minimization of Rational Functions on Compact Sets

We may generalize the previous example to rational objective functions. In particular, we now consider the global optimization problem:

$$val = \min_{x \in K} \frac{p(x)}{q(x)}, \qquad (9)$$

where p, q are polynomials such that $q(x) > 0 \; \forall \, x \in K$, and $K \subseteq \mathbb{R}^n$ is compact.

This problem has applications in many areas, including signal recovery [5] and finding minimal energy configurations of point charges in a field with polynomial potential [53].

It is simple to see that we may reformulate this problem as the GPM with $m = 1$ and $f_0 = p$, $f_1 = q$, and $b_1 = 1$, i.e.:

$$val = \min_{\mu \in M(K)_+} \left\{ \int_K p(x) d\mu(x) \; : \; \int_K q(x) d\mu(x) = 1 \right\}.$$

Indeed, one may readily verify that if x^* is a global minimizer of the rational function $p(x)/q(x)$ over K then an optimal solution of the GPM is given by $\mu^* = \frac{1}{q(x^*)} \delta_{x^*}$.

Polynomial Cubature Rules

Positive cubature (also known as multivariate quadrature) rules for numerical integration of a function f with respect to a measure μ_0 over a set K take the form

$$\int_K f(x)d\mu_0(x) \approx \sum_{\ell=1}^{N} w_\ell f(x^{(\ell)}),$$

where the points $x^{(\ell)} \in K$ and the weights $w_\ell \geq 0$ ($\ell \in [N]$) are fixed. The points (also known as the nodes of the cubature rule) and weights are typically chosen so that the approximation is exact for polynomials up to a certain degree, say d.

The problem of finding the points $x^{(\ell)} \in K$ and weights w_ℓ ($\ell \in [N]$) giving a cubature rule exact at degree d may then be written as the following GPM:

$$val := \inf_{\mu \in \mathcal{M}(K)_+} \left\{ \int_K 1 d\mu(x) \;:\; \int_K x^\alpha d\mu(x) = \int_K x^\alpha d\mu_0(x) \; \forall \alpha \in \mathbb{N}_d^n \right\}.$$

The key observation is that, by Theorem 3, this problem has an atomic solution supported on at most $N = |\mathbb{N}_d^n| = \binom{n+d}{d}$ points in K, say $\mu^* = \sum_{\ell=1}^{N} w_\ell \delta_{x^{(\ell)}}$, and this yields the cubature weights and points. This result is known as Tchakaloff's theorem [58]; see also [3, 57]. (In fact, our running assumption that K is compact may be relaxed somewhat in Tchakaloff's theorem—see, e.g. [46]).

Here we have chosen the constant polynomial 1 as objective function so that the optimal value is $val = \mu_0(K)$. Other choices of objective functions are possible as discussed, e.g., in [49]. The GPM formulation of the cubature problem was used for the numerical calculation of cubature schemes for various sets K in [49].

3   Semidefinite Programming Reformulations of the Approximations

The inner and outer approximations of the cone $\mathcal{M}(K)_+$ discussed in Sect. 1.4 lead to upper and lower bounds for the GPM (1), which may be reformulated as finite-dimensional, convex optimization problems, namely semidefinite programming (SDP) problems. These are conic linear programs over the cone of positive semidefinite matrices, formally defined as follows.

Semidefinite Programming (SDP) Problem

Assume we are given symmetric matrices $A_0, \ldots, A_m$ (all of the same size) and scalars $b_i \in \mathbb{R}$ ($i \in [m]$). The semidefinite programming problem in standard primal form is then defined as

$$p^* := \inf_{X \succeq 0} \{\langle A_0, X \rangle \; : \; \langle A_i, X \rangle = b_i \; \forall i \in [m]\},$$

where $\langle \cdot, \cdot \rangle$ now denotes the trace inner product, i.e., the Euclidean inner product in the space of symmetric matrices, and $X \succeq 0$ means that X is a symmetric positive semidefinite matrix (corresponding to the Löwner partial ordering of the symmetric matrices).

The dual semidefinite program reads

$$d^* := \sup_{y \in \mathbb{R}^m} \left\{ \sum_{i=1}^m b_i y_i \; : \; A_0 - \sum_{i=1}^m y_i A_i \succeq 0 \right\}.$$

Weak duality holds: $p^* \geq d^*$. Moreover, strong duality: $p^* = d^*$ holds, e.g., if the primal problem is bounded and admits a positive definite feasible solution X (or if the dual is bounded and has a feasible solution y for which $A_0 - \sum_i y_i A_i$ is positive definite) (see, e.g., [2, 4]).

Next we recall how one can test whether a polynomial can be written as a sum of squares of polynomials using semidefinite programming. This well known fact plays a key role for reformulating the inner and outer approximations of $\mathcal{M}(K)_+$ using semidefinite programs.

Checking Sums of Squares with SDP

Given an integer $r \in \mathbb{N}$ let $[x]_r = \{x^\alpha \; : \; \alpha \in \mathbb{N}_r^n\}$ consist of all monomials with degree at most r, thus the monomial basis of $\mathbb{R}[x]_r$.

Proposition 1 *For a given n-variate polynomial h, one has $h \in \Sigma_r$, if and only if the following polynomial identity holds:*

$$h(x) = [x]_r^\top M[x]_r \left(= \sum_{\alpha, \beta \in \mathbb{N}_r^n} M_{\alpha,\beta} x^{\alpha+\beta} \right),$$

for some positive semidefinite matrix: $M = \left(M_{\alpha,\beta} \right)_{\alpha,\beta \in \mathbb{N}_r^n} \succeq 0$. *The above identity can be equivalently written as*

$$h_\gamma = \sum_{\alpha, \beta \in \mathbb{N}_r^n \, : \, \alpha+\beta=\gamma} M_{\alpha,\beta} \quad \forall \gamma \in \mathbb{N}_{2r}^n. \tag{10}$$

Example 1

To illustrate the above algorithmic procedure for finding sums of squares, consider the following univariate polynomial

$$f(x) = 1 - 2x + 3x^2 - 2x^3 + x^4.$$

In order to check whether f can be written as a sums of squares we have to check the feasibility of the following semidefinite program, where the matrix variable M is a 3×3 symmetric matrix (indexed by the monomials $1, x, x^2$):

$$1 - 2x + 3x^2 - 2x^3 + x^4 = [x]_2^T M[x]_2, \ M \succeq 0.$$

By equating coefficients in the polynomials at both sides of the above identity we arrive at the following form for the matrix variable:

$$M_a = \begin{pmatrix} 1 & -1 & a \\ -1 & 3-2a & -1 \\ a & -1 & 1 \end{pmatrix} \quad \text{for some scalar } a.$$

One can check that the matrix M_a is positive semidefinite if and only if a satisfies $-1/2 \leq a \leq 1$. Hence, any value a in this interval provides a sum of squares decomposition for the polynomial f. For instance, the values $a = 1$ and $a = -1/2$ provide, respectively, the following factorizations for the matrix M_a:

$$M_1 = \begin{pmatrix} 1 \\ -1 \\ 1 \end{pmatrix} \begin{pmatrix} 1 \\ -1 \\ 1 \end{pmatrix}^T \quad \text{and} \quad M_{-1/2} = \frac{3}{4} \begin{pmatrix} 1 \\ 0 \\ -1 \end{pmatrix} \begin{pmatrix} 1 \\ 0 \\ -1 \end{pmatrix}^T + \frac{1}{4} \begin{pmatrix} 1 \\ -4 \\ 1 \end{pmatrix} \begin{pmatrix} 1 \\ -4 \\ 1 \end{pmatrix}^T ,$$

which in turn correspond to the following two decompositions of the polynomial f, respectively, as a single square and as a sum of two squares:

$$f(x) = (1 - x + x^2)^2 \quad \text{and} \quad f(x) = \frac{3}{4}(x - x^2)^2 + \frac{1}{4}(x - 4x + x^2)^2.$$

Note that, for any scalar a such that $-1/2 < a < 1$, the matrix M_a is positive definite and thus it provides a decomposition of the polynomial f as a sum of three squares.

Example 2

The Motzkin polynomial,

$$h(x_1, x_2) = x_1^4 x_2^2 + x_1^2 x_2^4 - 3x_1^2 x_2^2 + 1, \tag{11}$$

is nonnegative on $\mathbb{R}^2$ with roots at $(\pm 1, \pm 1)$ (see Fig. 1), but it is not a sum-of-squares of polynomials. It is an instructive exercise to show that the Motzkin polynomial does not satisfy the relations (10) for any $M = \left(M_{\alpha,\beta}\right)_{\alpha,\beta \in \mathbb{N}_3^n} \succeq 0$. For more details on the history of the Motzkin polynomial, see [47].

SDP Upper Bounds for GPM via the Inner Approximations
Recall that the inner approximations of the cone $\mathcal{M}(K)_+$ restrict the measures on K to the subsets $\mathcal{M}_{\mu_0}^r$ in (5), i.e. to those measures μ of the form $d\mu = h \cdot d\mu_0$, where μ_0 is a fixed reference measure with $\mathrm{Supp}(\mu_0) = K$ and $h \in \Sigma_r$ is a sum-of-squares polynomial density.

Replacing the cone $\mathcal{M}(K)_+$ in the GPM (1) by its subcone $\mathcal{M}_{\mu_0}^r$ we obtain the parameter

$$val_{inner}^{(r)} := \inf_{\mu \in \mathcal{M}_{\mu_0}^r} \left\{ \int_K f_0(x) d\mu(x) : \int_K f_i(x) d\mu(x) = b_i \ \forall i \in [m] \right\}, \tag{12}$$

which provides a hierarchy of upper bounds for GPM:

$$val \leq val_{inner}^{(r+1)} \leq val_{inner}^{(r)}.$$

Fig. 1 Plot of the Motzkin polynomial

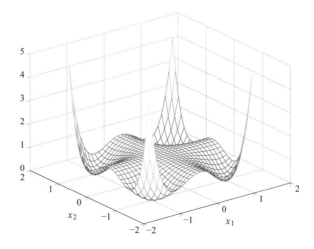

According to the above discussion these parameters can be reformulated as semidefinite programs involving the moments of the reference measure μ_0. Indeed, we may write the variable density function as $h(x) = [x]_r^T M[x]_r$ with $M \succeq 0$ and arrive at the following semidefinite program (in standard primal form).

SDP Formulation for the Inner Approximations Based Upper Bounds

$$val_{inner}^{(r)} = \inf_M \left\{ \langle A_0, M \rangle : \langle A_i, M \rangle = b_i \; \forall i \in [m], \; M = (M_{\alpha,\beta})_{\alpha,\beta \in \mathbb{N}_r^n} \succeq 0 \right\},$$

$$(13)$$

where we set

$$A_i = \int_K f_i(x)[x]_r[x]_r^T d\mu_0(x) = \left(\int_K f_i(x) x^{\alpha+\beta} d\mu_0(x) \right)_{\alpha,\beta \in \mathbb{N}_r^n} \quad (0 \leq i \leq m).$$

Moreover, writing each polynomial f_i in the monomial basis as $f_i = \sum_\gamma f_{i,\gamma} x^\gamma$ one sees that the entries of the matrix A_i depend linearly on the moments of the reference measure μ_0, since $\int_K f_i(x) x^{\alpha+\beta} d\mu_0(x) = \sum_\gamma f_{i,\gamma} m_{\alpha+\beta+\gamma}^{\mu_0}(K)$.

To be able to compute the above SDP one needs the moments of the reference measure μ_0 to be known on the set K. This is a restrictive assumption, since even computing volumes of polytopes is an NP-hard problem. One is therefore restricted to specific choices of μ_0 and K where the moments are known in closed form (or can be derived). In Table 1 we therefore give an overview of some known moments for the Euclidean ball and sphere, the hypercube, and the standard simplex. (See [25] for an easy derivation of the moments on the ball and the sphere.) There we use the Gamma function:

$$\Gamma(k) = (k-1)!, \quad \Gamma\left(k + \frac{1}{2}\right) = \left(k - \frac{1}{2}\right)\left(k - 1 - \frac{1}{2}\right)\cdots\frac{1}{2}\sqrt{\pi} \quad \text{for } k \in \mathbb{N}.$$

Table 1 Examples of known moments for some choices of $K \subseteq \mathbb{R}^n$: $\Delta_n = \{x \in \mathbb{R}_+^n : \sum_{i=1}^n x_i = 1\}$ is the standard simplex and $B_n = \{x \in \mathbb{R}^n : \|x\| \leq 1\}$ is the unit Euclidean ball, in which case μ_0 is the Lebesgue measure, and $S_n = \{x \in \mathbb{R}^n : \|x\| = 1\}$ is the unit Euclidean sphere in which case μ_0 is the (Haar) surface measure on S_n

K	$m_\alpha^{\mu_0}(K)$
$[0,1]^n$	$\prod_{i=1}^n \frac{1}{\alpha_i + 1}$
Δ_n	$\frac{\prod_{i=1}^n \alpha_i!}{(\sum_{i=1}^n \alpha_i + n)!}$
S_n	$\begin{cases} \frac{2\Gamma(\beta_1)\cdots\Gamma(\beta_n)}{\Gamma(\beta_1+\ldots+\beta_n)} & \text{if } \alpha \in (2\mathbb{N})^n \quad \text{with } \beta_i = \frac{\alpha_i+1}{2} \text{ for } i \in [n] \\ 0 & \text{otherwise} \end{cases}$
B_n	$\begin{cases} \frac{1}{\alpha_1+\ldots+\alpha_n+n} \frac{2\Gamma(\beta_1)\cdots\Gamma(\beta_n)}{\Gamma(\beta_1+\ldots+\beta_n)} & \text{if } \alpha \in (2\mathbb{N})^n \quad \text{with } \beta_i = \frac{\alpha_i+1}{2} \text{ for } i \in [n] \\ 0 & \text{otherwise} \end{cases}$

If K is an ellipsoid, one may obtain the moments of the Lebesgue measure on K from the moments on the ball by an affine transformation of variables. Also, if K is a polytope, one may obtain the moments of the Lebesgue measure through triangulation of K, and subsequently using the formula for the simplex.

Example 2 (Continued)

As an example we illustrate the inner approximation hierarchy for the problem of minimizing the Motzkin polynomial (11) on $[-2, 2]^2$ with the Lebesgue measure as reference measure. In Fig. 2, we plot the optimal density functions $h \in \Sigma_r$ for $r = 6, 8, 10, 12$. Note that, as r grows, the density functions become increasingly better approximations of a convex combination of the four the Dirac delta measures, centered at $(\pm 1, \pm 1)$. The corresponding upper bounds are $val_{inner}^{(6)} = 0.801069$, $val_{inner}^{(8)} = 0.565553$, $val_{inner}^{(10)} = 0.507829$, and $val_{inner}^{(12)} = 0.406076$. Note that these upper bounds are monotonically decreasing with increasing r, and recall that the minimum value of the Motzkin polynomial is zero.

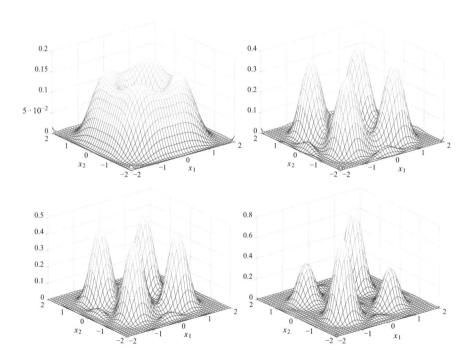

Fig. 2 Plots of the optimal density functions $h \in \Sigma_r$ for $r = 6, 8, 10, 12$

SDP Lower Bounds for GPM via the Outer Approximations
Here we assume that K is basic closed semi-algebraic, of the form

$$K = \{x \in \mathbb{R}^n : g_j(x) \geq 0 \ \forall j \in [k]\}, \quad \text{where } g_1, \dots, g_k \in \mathbb{R}[x].$$

Recall that the dual cone of the truncated quadratic module generated by the polynomials g_j describing the set K provides an outer approximation of $M(K)_+$; we repeat its definition (7) for convenience:

$$(Q^r(g_1, \dots, g_k))^* = \left\{ \mu \in M(K) : \int_K f d\mu \geq 0 \ \forall f \in Q^r(g_1, \dots, g_k) \right\},$$

where the quadratic module $Q^r(g_1, \dots, g_k)$ was defined in (6).

Replacing the cone $M(K)_+$ in the GPM (1) by the above outer approximations we obtain the following parameters

$$val_{outer}^{(r)} := \inf_{\mu \in (Q^r(g_1, \dots, g_k))^*} \left\{ \int_K f_0(x) d\mu(x) : \int_K f_i(x) d\mu(x) = b_i \ \forall i \in [m] \right\}, \tag{14}$$

which provide a hierarchy of lower bounds for the GPM:

$$val_{outer}^{(r)} \leq val_{outer}^{(r+1)} \leq val.$$

Here too these parameters can be reformulated as semidefinite programs. Indeed a signed measure μ lies in the cone $(Q^r(g_1, \dots, g_k))^*$ precisely when it satisfies the condition

$$\int_K g_j(x) \sigma_j(x) d\mu(x) \geq 0 \ \forall \sigma_j \in \Sigma_{r_j}, \quad \forall j \in \{0, \dots, k\}, \tag{15}$$

where $r_j = r - \lceil \deg(g_j)/2 \rceil$. Using Proposition 1, we may represent each sum-of-squares σ_j as

$$\sigma_j(x) = [x]_{r_j}^{\mathsf{T}} M^{(j)} [x]_{r_j}$$

for some matrix $M^{(j)} \succeq 0$ (indexed by $\mathbb{N}_{r_j}^n$). Hence we have

$$\int_K g_j(x) \sigma_j(x) d\mu(x) = \int_K g_j(x) [x]_{r_j}^{\mathsf{T}} M^{(j)} [x]_{r_j} d\mu(x) = \langle B_j^\mu, M^{(j)} \rangle,$$

after setting

$$B_j^\mu = \int_K g_j(x) [x]_{r_j} [x]_{r_j}^{\mathsf{T}} d\mu(x) = \left(\int_K g_j(x) x^{\alpha+\beta} d\mu(x) \right)_{\alpha, \beta \in \mathbb{N}_{r_j}^n}.$$

Hence the condition (15) can be rewritten as requiring, for each $j \in \{0, 1, \ldots, k\}$,

$$\langle B_j^\mu, M^{(j)} \rangle \geq 0 \quad \text{for all postive semidefinite matrices } M^{(j)} \text{ indexed by } \mathbb{N}_{r_j}^n,$$

which in turn is equivalent to $B_j^\mu \succeq 0$ (since the cone of positive semidefinite matrices is self-dual). Summarizing, the condition (15) on the variable measure μ can be rewritten as

$$B_j^\mu = \left(\int_K g_j(x) x^{\alpha+\beta} d\mu(x) \right)_{\alpha, \beta \in \mathbb{N}_{r_j}^n} \succeq 0 \quad \forall j \in \{0, 1, \ldots, k\}.$$

Finally, observe that only the moments of μ are playing a role in the above constraints. Therefore we may introduce new variables for these moments, say

$$y_\alpha = \int_K x^\alpha d\mu(x) \quad \forall \alpha \in \mathbb{N}_{2r}^n.$$

Writing the polynomials g_j in the monomial basis as $g_j(x) = \sum_\gamma g_{j,\gamma} x^\gamma$ we arrive at the following SDP reformulation for the parameter $val_{outer}^{(r)}$.

SDP Formulation for the Outer Approximations Based Lower Bounds

With $r_j = r - \lceil \deg(g_j)/2 \rceil$ for $j \in \{0, 1, \ldots, k\}$ and d an upper bound on the degrees of f_i for $i \in \{0, 1, \ldots, m\}$ we have

$$val_{outer}^{(r)} = \inf_{(y_\alpha)_{\alpha \in \mathbb{N}_{2r}^n}} \left\{ \sum_{\alpha \in \mathbb{N}_d^n} f_{0,\alpha} y_\alpha : \sum_{\alpha \in \mathbb{N}_d^n} f_{i,\alpha} y_\alpha = b_i \quad \forall i \in [m], \quad (16) \right.$$

$$\left. \left(\sum_\gamma g_{j,\gamma} y_{\alpha+\beta+\gamma} \right)_{\alpha, \beta \in \mathbb{N}_{r_j}^n} \succeq 0 \quad \forall j \in \{0, 1, \ldots, k\} \right\}. \quad (17)$$

Example 2 (Continued)

We now illustrate the hierarchy of outer approximations for the minimization of the Motzkin polynomial (11) on $K = [-2, 2]^2$. If we represent K by the linear inequalities

$$-2 \leq x_1 \leq 2, \quad -2 \leq x_2 \leq 2,$$

then the lower bounds on the zero minimum become

$$val_{outer}^{(3)} = -1.6858, \quad val_{outer}^{(4)} = 0.$$

In other words, one has convergence in a finite number of steps here, namely already for $r = 4$. If one represents K by the quadratic inequalities

$$x_1^2 \leq 4, \ x_2^2 \leq 4,$$

then the convergence is even faster, since one then has $val_{outer}^{(3)} = 0$. It is therefore interesting to note that the description of K plays an important role for the outer approximations.

If, in the definition (14) of $val_{outer}^{(r)}$, instead of the truncated quadratic module $Q^r(g_1, \ldots, g_k)$ we use the larger quadratic module $Q^r(\prod_{j \in J} g_j : J \subseteq [k])$ generated by the pairwise products of the g_j's, then we obtain a stronger bound on val, which we denote by $\overline{val}_{outer}^{(r)}$. Thus

$$\overline{val}_{outer}^{(r)} = \inf_{\mu \in (Q^r(\prod_{j \in J} g_j : J \subseteq [k]))^*} \left\{ \int_K f_0(x) d\mu(x) : \int_K f_i(x) d\mu(x) = b_i \ (i \in [m]) \right\}$$
(18)

and clearly we have

$$val_{outer}^{(r)} \leq \overline{val}_{outer}^{(r)} \leq val.$$

The parameter $\overline{val}_{outer}^{(r)}$ can also be reformulated as a semidefinite program, analogous to the program (16)–(17), which however now involves $2^k + 1$ semidefinite constraints instead of $k + 1$ such constraints in (17) and thus its practical implementation is feasible only for small values of k. On the other hand, as we will see later in Sect. 5.2, the bounds $\overline{val}_{outer}^{(r)}$ admit a much sharper error analysis than the bounds $val_{outer}^{(r)}$ for the case of polynomial optimization.

4 Convergence Results for the Inner Approximation Hierarchy

In the rest of the paper we are interested in the convergence of the respective lower and upper SDP bounds on the optimal value of the GPM, as introduced in the previous section. We will first consider in this section the upper bounds for the GPM arising from the inner approximations, since much more is known about their rate of convergence than for the lower bounds arising from the outer approximations. We deal first with the special case of polynomial optimization and then indicate how some of the results extend to the general GPM.

4.1   The Special Case of Global Polynomial Optimization

Here we consider a special case of the GPM, namely global optimization of polynomials on compact sets (i.e., problem (8)) and review the main known results about the error analysis of the upper bounds $val^{(r)}_{inner}$. After that in the next section we will explain how to extend this error analysis to the bounds for the general GPM problem.

Thus we now consider the problem

$$val = \min_{x \in K} p(x), \tag{19}$$

asking to find the minimum value of the polynomial $p(x) = \sum_{\alpha \in \mathbb{N}^n_d} p_\alpha x^\alpha$ over a compact set K.

Recall the definition of the inner approximation based upper bound (12), which can be rewritten here as

$$val^{(r)}_{inner} = \min_{h \in \Sigma_r} \left\{ \int_K p(x)h(x)d\mu_0(x) \; : \; \int_K h(x)d\mu_0(x) = 1 \right\},$$

and its SDP reformulation from (13), which now reads

$$val^{(r)}_{inner} = \min \left\{ \langle A_0, M \rangle : \langle A_1, M \rangle = 1, \; M = (M_{\alpha,\beta})_{\alpha,\beta \in \mathbb{N}^n_r} \succeq 0 \right\}, \tag{20}$$

with

$$A_0 = \left(\int_K p(x)x^{\alpha+\beta}d\mu_0(x) \right)_{\alpha,\beta \in \mathbb{N}^n_r}, \quad A_1 = \left(\int_K x^{\alpha+\beta}d\mu_0(x) \right)_{\alpha,\beta \in \mathbb{N}^n_r},$$

where as before μ_0 is a fixed reference measure on K.

A first observation made in [35] is that this semidefinite program (20) can in fact be reformulated as a generalized eigenvalue problem. Indeed, its dual semidefinite program reads

$$\max\{\lambda : A_0 - \lambda A_1 \succeq 0\},$$

whose optimal value gives again the parameter $val^{(r)}_{inner}$ (since strong duality holds). Hence $val^{(r)}_{inner}$ is equal to the smallest generalized eigenvalue of the system

$$A_0 v = \lambda A_1 v, \quad v \neq 0. \tag{21}$$

Thus one may compute $val^{(r)}_{inner}$ without having to solve an SDP problem.

In fact, if instead of the monomial basis $\{x^\alpha : \alpha \in \mathbb{R}^n_{2r}\}$ we use a polynomial basis $\{b_\alpha(x) : \alpha \in \mathbb{N}^n_{2r}\}$ of $\mathbb{R}[x]_{2r}$ that is orthonormal with respect to the reference measure μ_0 (i.e., such that $\int_K b_\alpha b_\beta d\mu_0 = 1$ if $\alpha = \beta$ and 0 otherwise), then in the above semidefinite program (20) we may set $A_1 = I$ to be the identity matrix and

$$A_0 = \left(\int_K p(x)b_\alpha(x)b_\beta(x)d\mu_0(x) \right)_{\alpha,\beta\in\mathbb{N}_{2r}^n}, \tag{22}$$

whose entries now involve the 'generalized' moments $\int_K b_\alpha(x)d\mu_0(x)$ of μ_0. Then the parameter $val_{inner}^{(r)}$ can be computed as the smallest eigenvalue of the matrix A_0:

$$val_{inner}^{(r)} = \lambda_{\min}(A_0) \quad \text{where } A_0 \text{ is as in (22).} \tag{23}$$

This fact was observed in [14] and used there to establish a link with the roots of the orthonormal polynomials, permitting to analyze the quality of the bounds $val_{inner}^{(r)}$ for the case of the hypercube $K = [-1, 1]^n$, see below for details.

In Table 2 we list the known convergence rates of the parameters $val_{inner}^{(r)}$ to the optimal value val of problem (19), i.e., we review the known upper bounds for the sequence $\{val_{inner}^{(r)} - val\}, r = 1, 2, \ldots$

We will give some details on the proofs of each of the four results listed in Table 2. After that we will mention an interesting connection with approximations based on cubature rules.

Asymptotic Convergence

The first result in Table 2 states that $\lim_{r\to\infty} val_{inner}^{(r)} = val$ if K is compact and $\mu_0 \in M(K)_+$. It is a direct consequence of the following result.

Theorem 4 (Lasserre [35]) *Let $K \subseteq \mathbb{R}^n$ be compact, let μ_0 be a fixed, finite, positive Borel measure with $\text{Supp}(\mu_0) = K$. and let f be a continuous function on $\mathbb{R}^n$. Then, f is nonnegative on K if and only if*

$$\int_K g^2 f d\mu_0 \geq 0 \ \forall g \in \mathbb{R}[x].$$

Table 2 Known rates of convergence for the Lasserre hierarchy of upper bounds on val in (19) based on inner approximations

$K \subseteq \mathbb{R}^n$	$val_{inner}^{(r)} - val$	Measure μ_0	Reference
Compact	$o(1)$	Positive finite Borel measure	[35]
Compact, satisfies interior cone condition	$O\left(\frac{1}{\sqrt{r}}\right)$	Lebesgue measure	[18]
Convex body	$O\left(\frac{1}{r}\right)$	Lebesgue measure	[13]
Hypercube $[-1, 1]^n$	$\Theta\left(\frac{1}{r^2}\right)$	$\prod_{i=1}^n (1 - x_i^2)^{-1/2}dx_i$	[14]
Unit sphere, p homogeneous	$O\left(\frac{1}{r}\right)$	Surface measure	[21]

The asymptotic convergence of the bounds $val^{(r)}_{inner}$ to val holds more generally for the minimization of a rational function $p(x)/q(x)$ over K (assuming $q(x) > 0$ for all $x \in K$). Indeed, using the above theorem, we obtain

$$\min_{x \in K} \frac{p(x)}{q(x)} = \sup_{t \in \mathbb{R}} t \;\; \text{s.t.} \;\; p(x) \geq tq(x) \;\; \forall x \in K$$

$$= \sup_{t \in \mathbb{R}} t \;\; \text{s.t.} \;\; \int_K p(x)h(x)d\mu_0(x) \geq t \int_K q(x)h(x)d\mu_0(x) \;\; \forall h \in \Sigma$$

$$= \inf_{h \in \Sigma} \int_K p(x)h(x)d\mu_0(x) \;\; \text{s.t.} \;\; \int_K q(x)h(x)d\mu_0(x) = 1.$$

Error Analysis When K Is Compact and Satisfies an Interior Cone Condition
The second result in Table 2 fixes the reference measure μ_0 to the Lebesgue measure, and restricts the set K to satisfy a so-called interior cone condition.

Definition 1 (Interior Cone Condition) A set $K \subseteq \mathbb{R}^n$ satisfies an interior cone condition if there exist an angle $\theta \in (0, \pi/2)$ and a radius $\rho > 0$ such that, for every $x \in K$, a unit vector $\xi(x)$ exists such that

$$\{x + \lambda y : y \in \mathbb{R}^n, \|y\| = 1, y^T \xi(x) \geq \cos\theta, \lambda \in [0, \rho]\} \subseteq K.$$

For example, all full-dimensional convex sets satisfy the interior cone condition for suitable parameters θ and ρ. This assumption is used in [18] to claim that the intersection of any ball with the set K contains a positive fraction of the full ball, a fact used in the error analysis.

The main ingredient of the proof is to approximate the Dirac delta supported on a global minimizer by a Gaussian density of the form

$$G(x) = \frac{1}{(2\pi\sigma^2)^{n/2}} \exp\left(\frac{-\|x - x^*\|^2}{2\sigma^2} \right), \tag{24}$$

where x^* is a minimizer of p on K, and $\sigma^2 = \Theta(1/r)$. Then we approximate the Gaussian density $G(x)$ by a sum-of-squares polynomial $g_r(x)$ with degree $2r$. For this we use the fact that the Taylor approximation of the exponential function e^{-t} is a sum of squares (since it is a univariate polynomial nonnegative on $\mathbb{R}$).

Lemma 1 *For any $r \in \mathbb{N}$ the univariate polynomial $\sum_{k=0}^{2r} \frac{(-1)^k}{k!} t^k$ (in the variable $t \in \mathbb{R}$), defined as the Taylor expansion of the function $t \in \mathbb{R} \mapsto e^{-t}$ truncated at degree $2r$, is a sum of squares of polynomials.*

Based on this the polynomial

$$g_r(x) = \frac{1}{(2\pi\sigma^2)^{n/2}} \sum_{k=0}^{2r} \frac{(-1)^k}{k!} \left(\frac{-\|x - x^*\|^2}{2\sigma^2} \right)^k$$

is indeed a sum of squares with degree $2r$, which can be used (after scaling) as feasible solution within the definition of the bound $val^{(r)}_{inner}$. We refer to [18] for the details of the analysis.

Error Analysis When K Is a Convex Body
The third item in Table 2 assumes that K is now convex, compact and full-dimensional, i.e., a convex body. The key idea is to use the following concentration result for the *Boltzman density* (or *Gibbs measure*).[3]

Theorem 5 (Kalai-Vempala [28]) *If p is a linear polynomial, K is a convex set, $T > 0$ is a fixed 'temperature' parameter, and $val = \min_{x \in K} p(x)$, then we have*

$$\int_{\mathbf{K}} p(x) H(x) dx - val \leq nT,$$

where

$$H(x) = \frac{\exp(-p(x)/T)}{\int_K \exp(-p(x)/T) dx}$$

is the Boltzman probability density supported on K.

The theorem still holds if p is convex, but not necessarily linear [13]. The proof of the third item in Table 2 now proceeds as follows:

1. Construct a sum-of-squares polynomial approximation $h_r(x)$ of the Boltzman density $H(x)$ by again using the fact that the even degree truncated Taylor expansion of e^{-t} is a sum of squares (Lemma 1); namely, consider the polynomial $h_r(x) = \sum_{k=0}^{2r} \frac{(-1)^k}{k!} \left(\frac{-p(x)}{T}\right)^k$ (up to scaling).
2. Use this construction to bound the difference between $val^{(r)}_{inner}$ and the Boltzman bound when choosing $T = O(1/r)$.
3. Use the extension of the Kalai-Vempala result to get the required result for convex polynomials p.
4. When p is nonconvex, the key ingredient is to reduce to the convex case by constructing a convex (quadratic) polynomial $\hat{p}$ that upper bounds p on K and has the same minimizer on K, as indicated in the next lemma.

Lemma 2 *Assume x^* is a global minimizer of p over K. Then the following polynomial*

$$\hat{p}(x) = p(x^*) + \nabla p(x^*)^T (x - x^*) + C_p \|x - x^*\|^2$$

with $C_p = \max_{x \in K} \|\nabla^2 p(x)\|_2$, is quadratic, convex, and separable. Moreover, it satisfies: $p(x) \leq \hat{p}(x)$ for all $x \in K$, and x^* is a global minimizer of $\hat{p}$ over K.

---

[3]This result is of independent interest in the study of *simulated annealing* algorithms.

Then, in view of the inequality

$$\int_K \hat{p} h \, d\mu_0 \geq \int_K p h \, d\mu_0 \quad \forall h \in \Sigma_r, \tag{25}$$

it follows that the error analysis in the non-convex case follows directly from the error analysis in the convex case. The details of the proof are given in [13].

Error Analysis for the Hypercube $K = [-1, 1]^n$
The fourth result in Table 2 deals with the hypercube $K = [-1, 1]^n$. A first key idea of the proof is that it suffices to show the $O(1/r^2)$ convergence rate for a univariate quadratic polynomial. This follows from Lemma 2 above (and (25)), which implies that it suffices to analyze the case of a quadratic, separable polynomial. Hence we may further restrict to the case when $K = [-1, 1]$ and p is a quadratic univariate polynomial.

In the univariate case, the key idea is to use the eigenvalue reformulation of the bound $val_{inner}^{(r)}$ from (23). There, we use the polynomial basis $\{b_k : k \in \mathbb{N}\}$ consisting of the Chebyshev polynomials (of the first kind) which are orthonormal with respect to the Chebyshev measure $d\mu_0$ on $K = [-1, 1]$, indeed the measure used in Table 2.

Then one may use a connection to the extremal roots of these orthonormal polynomials. Namely, for the linear polynomial $p(x) = x$, the parameter $val_{inner}^{(r)}$ coincides with the smallest root of the orthonormal polynomial b_{r+1} (with degree $r + 1$); this is a well known property of orthogonal polynomials, which follows from the fact that the matrix A_0 in (22) is tri-diagonal and the 3-terms recurrence for the Chebyshev polynomials (see, e.g., [22, §1.3]). When p is a quadratic polynomial, the matrix A_0 in the eigenvalue problem (23) is now 5-diagonal and 'almost' Toepliz, properties that can be exploited to evaluate its smallest eigenvalue. See [14] for details.

Error Analysis for the Unit Sphere
The last result in Table 2 deals with the minimization of a homogeneous polynomial p over the unit sphere $S_n = \{x \in \mathbb{R}^n : \sum_{i=1}^n x_i^2 = 1\}$, in which case Doherty and Wehner [21] show a convergence rate in $O(1/r)$. Their construction for a suitable sum-of-squares polynomial density in Σ_r is in fact closely related to their analysis of the outer approximation based lower bounds $val_{outer}^{(r)}$. Doherty and Wehner [21] indeed show the following stronger result: $val_{inner}^{(r)} - val_{outer}^{(r)} = O(1/r)$, to which we will come back in Sect. 5.2 below.

Link with Positive Cubature Rules
There is an interesting link between positive cubature formulas and the upper bound

$$val_{inner}^{(r)} = \min_{h \in \Sigma_r} \left\{ \int_K p h \, d\mu_0 : \int_K h \, d\mu_0 = 1 \right\},$$

which was recently pointed out in [39] and is summarized in the next result.

Theorem 6 (Martinez et al. [39]) *Let* $x^{(1)}, \ldots, x^{(N)} \in K$ *and weights* $w_1 > 0, \ldots, w_N > 0$ *give a positive cubature rule on* K *for the measure* μ_0, *that is exact for polynomials of total degree at most* $d + 2r$, *where* $d > 0$ *and* $r > 0$ *are given integers. Let* p *be a polynomial of degree* d.

Then, if h *is a polynomial nonnegative on* K *and of degree at most* $2r$, *one has*

$$\int_K p h \, d\mu_0 \geq \min_{\ell \in [N]} p(x^{(\ell)}).$$

In particular, the inner approximation bounds therefore satisfy

$$val_{inner}^{(r)} \geq \min_{\ell \in [N]} p(x^{(\ell)}).$$

The proof is an immediate consequence of the definitions, but this result has several interesting implications.

- First of all, one may derive information about the rate of convergence for the scheme $\min_{\ell \in [N]} p(x^{(\ell)})$ from the error bounds in Table 2. For example, if K is a convex body, the implication is that $\min_{\ell \in [N]} p(x^{(\ell)}) - val = O(1/r)$.
- Also, if a positive cubature rule is known for the pair (K, μ_0), and the number of points N meets the Tchakaloff bound $N = \binom{n+2r+d}{2r+d}$, then there is no point in computing the parameter $val_{inner}^{(r)}$. Indeed, as

$$val_{inner}^{(r)} \geq \min_{\ell \in [N]} p(x^{(\ell)}) \geq val,$$

the right-hand-side bound is stronger and can be computed more efficiently. Having said that, positive cubature rules that meet the Tchakaloff bound are only known in special cases, typically in low dimension and degree; see e.g. [6, 8, 57], and the references therein.

- Theorem 6 also shows why the last convergence rate in Table 2 is tight for $K = [-1, 1]^n$. Indeed if we consider the univariate example $p(x) = x$ and the Chebyshev probability measure $d\mu_0(x) = \dfrac{1}{\pi\sqrt{1-x^2}} dx$ on $K = [-1, 1]$, then a positive cubature scheme is given by

$$x^{(\ell)} = \cos\left(\frac{2\ell - 1}{2N}\pi\right), \quad w_\ell = \frac{1}{N} \quad \forall \ell \in [N],$$

and it is exact at degree $2N - 1$. This is known as the Chebyshev-Gauss quadrature, and the points are precisely the roots of the degree N Chebyshev polynomial of the first kind. Thus, with $N = r + 1$, in this case we have

$$val_{inner}^{(r)} \geq \min_{\ell \in [N]} p(x^{(\ell)}) = \min_{\ell \in [N]} \cos\left(\frac{(2\ell - 1)\pi}{2N}\right) = \cos\left(-\pi/(2N)\right) = -1 + \Omega\left(\frac{1}{N^2}\right).$$

This explains that the $\Theta(1/r^2)$ result in Table 2 holds for $p(x) = x$. A different proof of this result is given in [14], where it is shown that for this example one actually has equality $val_{inner}^{(r)} = \cos(-\pi/(2N))$.

- Finally, Theorem 6 shows that there is not much gain in using a set of densities larger than Σ_r in the definition of the inner approximations $\mathcal{M}_{\mu_0}^r$ since the statement of the theorem holds for any nonnegative polynomial h on K. For example, for the hypercube $K = [-1, 1]^n$, if we use the larger set of densities $h \in Q^r(\prod_{j \in J}(1 - x_j^2) : J \subseteq [k])$ and the Chebyshev measure as reference measure μ_0 on $[-1, 1]^n$, then we obtain upper bounds with convergence rate in $O(1/r^2)$ [9]. This also follows from the later results in [14] where in addition it is shown that this convergence result is tight for linear polynomials. By the above discussion tightness also follows from Theorem 6.

Upper Bounds Using Grid Point Sets

Of course one may also obtain upper bounds on val, the minimum value taken by a polynomial p over a compact set K, by evaluating p at any suitably selected set of points in K. This corresponds to restricting the optimization over selected finite atomic measures in the definition of val.

A first basic idea is to select the grid point sets consisting of all rational points in K with denominator r for increasing values of $r \in \mathbb{N}$. For the standard simplex $K = \Delta_n$ and the hypercube $K = [0, 1]^n$ this leads to upper bounds that satisfy:

$$\min_{x \in K, rx \in \mathbb{N}^n} p(x) - \min_{x \in K} p(x) \leq \frac{C_d}{r}\left(\max_{x \in K} p(x) - \min_{x \in K} p(x)\right) \quad \text{for all } r \geq d,$$
(26)

where C_d is a constant that depends only on the degree d of p; see [17] for $K = \Delta_n$ and [12] for $K = [0, 1]^n$. A faster regime in $O(1/r^2)$ can be shown when allowing a constant that depends on the polynomial p (see [19] for Δ_n and [11] for $[0, 1]^n$). Note that the number of rational points with denominator r in the simplex Δ_n is $\binom{n+r-1}{r} = O(n^r)$ and thus the computation time for these upper bounds is polynomial in the dimension n for any fixed order r. On the other hand, there are $(r+1)^n = O(r^n)$ such grid points in the hypercube $[0, 1]^n$ and thus the computation time of the upper bounds grows exponentially with the dimension n.

For a general convex body K some constructions are proposed recently in [44] for suitable grid point sets (so-called meshed norming sets) $X_d(\epsilon) \subseteq K$ where $d \in \mathbb{N}$ and $\epsilon > 0$. Namely, whenever p has degree at most d, by minimizing p over $X_d(\epsilon)$ one obtains an upper bound on the minimum of p over K satisfying

$$\min_{x \in X_d(\epsilon)} p(x) - \min_{x \in K} p(x) \leq \epsilon \left(\max_{x \in K} p(x) - \min_{x \in K} p(x)\right),$$

where the computation involves $|X_d(\epsilon)| = O\left(\left(\frac{d}{\sqrt{\epsilon}}\right)^{2n}\right)$ point evaluations, thus exponential in the dimension n for fixed precison ϵ.

In comparison, the computation of the upper bound $val_{outer}^{(r)}$ relies on a semidefinite program involving a matrix of size $\binom{n+r}{r} = O(n^r)$, which is polynomial in the dimension n for any fixed order r.

4.2   The General Problem of Moments (GPM)

One may extend the results of the last section to the inner approximations for the general GPM (1). In other words, we now consider the upper bounds (12) obtained using the inner approximations of the cone $\mathcal{M}(K)_+$, which we repeat for convenience:

$$val_{inner}^{(r)} = \inf_{h \in \Sigma_r} \left\{ \int_K f_0(x)h(x)d\mu_0(x) \; : \; \int_K f_i(x)h(x)d\mu_0(x) = b_i \quad \forall i \in [m] \right\}.$$

A first observation is that this program may not have a feasible solution, even if the GPM (1) does. For example, two constraints like

$$\int_0^1 x d\mu(x) = 0, \quad \int_0^1 d\mu(x) = 1$$

admit the Dirac measure $\mu = \delta_{\{0\}}$ as solution but they do not admit any solution of the form $d\mu = hdx$ with $h \in \Sigma_r$ for any $r \in \mathbb{N}$. Thus any convergence result must relax the equality constraints of the GPM (1) in some way, or involve additional assumptions.

We now indicate how one may use the convergence results of the last section to derive an error analysis for the inner approximations of the GPM when relaxing the equality constraints.

Theorem 7 (De Klerk-Postek-Kuhn [20]) *Assume that $f_0, \ldots, f_m$ are polynomials, K is compact and the GPM (1) has an optimal solution. Let $b_0 := val$ denote the optimal value of (1) and for any integer $r \in \mathbb{N}$ define the parameter*

$$\Delta^{(r)} := \min_{h \in \Sigma_r} \max_{i \in \{0,1,\ldots,m\}} \left| \int_K f_i(x)h(x)d\mu_0(x) - b_i \right|.$$

Then the following assertions hold:

(1) $\lim_{r \to \infty} \Delta^{(r)} = 0.$

(2) $\Delta^{(r)} = O\left(\frac{1}{r^{1/4}}\right)$ *if K satisfies an interior cone assumption and μ_0 is the Lebesgue measure;*

(3) $\Delta^{(r)} = O\left(\frac{1}{r^{1/2}}\right)$ *if* K *is a convex body and* μ_0 *is the Lebesgue measure;*

(4) $\Delta^{(r)} = O\left(\frac{1}{r}\right)$ *if* $K = [-1, 1]^n$ *and* $d\mu_0(x) = \prod_i (1 - x_i^2)^{-1/2} dx_i.$

We will derive this from the convergence results for global polynomial optimization in Table 2. By assumption, problem (1) has an optimal solution and by Theorem 3 we may assume it has an atomic optimal solution $\mu^* = \sum_\ell \lambda_\ell \delta_{x_\ell^*}$ with $\lambda_\ell > 0$ and $x_\ell^* \in K$. We now sketch the proof.

1. For each atom x_ℓ^* of the optimal measure μ^* consider the polynomial

$$p_\ell(x) = \sum_{i=0}^{m} \left(f_i(x) - f_i(x_\ell^*)\right)^2,$$

   whose minimum value over K is equal to 0 (attained at x_ℓ^*).
2. We apply the error analysis of the previous section to the problem of minimizing the polynomial p_ℓ over K. In particular, the asymptotic convergence of the upper bounds implies that for any given $\epsilon > 0$

$$\exists r \in \mathbb{N} \ \exists h_\ell \in \Sigma_r \ \text{s.t.} \int_K p_\ell(x) h_\ell(x) d\mu_0(x) \le \epsilon^2, \int_K h_\ell(x) d\mu_0(x) = 1$$

   and, therefore,

$$\int_K (f_i(x) - f_i(x_\ell^*))^2 h_\ell(x) d\mu_0(x) \le \epsilon^2 \ \forall i \in \{0, \dots, m\}. \tag{27}$$

3. Using the Jensen inequality, one obtains

$$\left| \int_K f_i(x) h_\ell(x) d\mu_0(x) - f_i(x_\ell^*) \right| = \left| \int_K (f_i(x) - f_i(x_\ell^*)) h_\ell(x) d\mu_0(x) \right| \le \epsilon$$

   for each $i \in \{0, \dots, m\}$.
4. We now consider the sum-of-squares density $h := \sum_\ell \lambda_\ell h_\ell \in \Sigma_r$. Then we have $b_i = \int_K f_i(x) d\mu^*(x) = \sum_\ell \lambda_\ell f_i(x_\ell^*)$ for each $i \in \{0, \dots, m\}$. Moreover, the above argument shows that for any $i \in \{0, \dots, m\}$

$$\left| \int_K f_i(x) h(x) d\mu_0(x) - b_i \right| = \left| \sum_\ell \lambda_\ell \left(\int_K f_i(x) h_\ell(x) d\mu_0(x) - f_i(x_\ell^*) \right) \right| \le \epsilon \mu^*(K)$$

   with $\mu^*(K) = \sum_\ell \lambda_\ell$. This shows that $\Delta^{(r)} \le \epsilon \mu^*(K)$ and thus the desired asymptotic result (1).

5. The additional three claims (2)–(4) follow in the same way using the results in Table 2. For instance, in case (1) when K satisfies an interior cone condition and μ_0 is the Lebesgue measure, we replace the estimate (27) by

$$\left| \int_K (f_i(x) - f_i(x_\ell^*))^2 h_\ell(x) d\mu_0(x) \right| = O\left(\frac{1}{\sqrt{r}}\right),$$

which leads to $\Delta^{(r)} = O\left(\frac{1}{r^{1/4}}\right)$ (since we 'lose a square root' when applying Jensen inequality).

We may also use the relation with positive cubature rules discussed in the previous section (Theorem 6) to obtain the following cubature-based approximations for the GPM (1).

Corollary 2 *Assume the GPM (1) admits an optimal solution and let d denote the maximum degree of the polynomials $f_0, \ldots, f_m$. For any integer $r \in \mathbb{N}$ assume we have a cubature rule for (K, μ_0) that is exact for degree $d + 2r$, consisting of the points $x^{(\ell)} \in K$ and weights $w_\ell > 0$ for $\ell \in [N]$, and define the parameter*

$$\Delta_{cub}^{(r)} := \min_v \max_{i \in \{0,1,\ldots,m\}} \left| \int_K f_i(x) dv - b_i \right|,$$

where in the outer minimization we minimize over all atomic measures v whose atoms all belong to the set $\{x^{(\ell)} : \ell \in [N]\}$. Then the following assertions hold:

(1) $\lim\limits_{r\to\infty} \Delta_{cub}^{(r)} = 0;$

(2) $\Delta_{cub}^{(r)} = O\left(\frac{1}{r^{1/4}}\right)$ *if K satisfies an interior cone assumption and μ_0 is the Lebesgue measure;*

(3) $\Delta_{cub}^{(r)} = O\left(\frac{1}{\sqrt{r}}\right)$ *if K is a convex body and μ_0 is the Lebesgue measure;*

1. $\Delta_{cub}^{(r)} = O\left(\frac{1}{r}\right)$ *if $K = [-1, 1]^n$ and $d\mu_0(x) = \prod_i (1 - x_i^2)^{-1/2} dx_i$.*

This result follows from Theorem 7. Indeed, for any polynomial $h \in \Sigma_r$, the polynomials $f_i h$ have degree at most $d + 2r$ so that using the cubature rule we obtain

$$\int_K f_i(x) h(x) d\mu_0(x) = \sum_{\ell=1}^N w_\ell f_i(x^{(\ell)}) h(x^{(\ell)}) = \int_K f_i(x) dv(x),$$

where v is the atomic measure with atoms $x^{(\ell)}$ and weights $\alpha_\ell := w_\ell h(x^{(\ell)})$ for $\ell \in [N]$. Therefore, the parameter $\Delta_{cub}^{(r)}$ in Corollary 2 is upper bounded by the parameter $\Delta^{(r)}$ in Theorem 7. The claims (1)-(4) now follow directly from the corresponding claims in Theorem 7.

Note that, for any fixed $r \in \mathbb{N}$, in order to find the best atomic measure ν in the definition of $\Delta_{cub}^{(r)}$ we need to find the best weights α_ℓ ($\ell \in [N]$) giving the measure $\nu = \sum_{\ell=1}^{N} \alpha_\ell \delta_{x^{(\ell)}}$. This can be done by solving the following linear program:

$$\Delta_{cub}^{(r)} = \min_{t, \alpha_\ell \in \mathbb{R}} t \ \text{ s.t. } \alpha_\ell \geq 0 \ (\ell \in [N]), \ \left| \sum_{\ell=1}^{N} \alpha_\ell f_i(x^{(\ell)}) - b_i \right| \leq t \ \forall i \in \{0, 1, \ldots, m\}.$$

(This is similar to an idea used in [49].)

5 Convergence Results for the Outer Approximations

In this last section we consider the convergence of the lower bounds for the GPM (1), that are obtained by using outer approximations for the cone of positive measures. We first mention properties dealing with asymptotic and finite convergence for the general GPM and after that we mention some known results on the error analysis in the special case of polynomial optimization.

Here we assume K is a compact semi-algebraic set, defined as before by

$$K = \{x \in \mathbb{R}^n : g_j(x) \geq 0 \quad \forall j \in [k]\},$$

where $g_1, \ldots, g_k \in \mathbb{R}[x]$. We will consider the following (*Archimedean*) condition:

$$\exists r \in \mathbb{N} \ \exists u \in Q^r(g_1, \ldots, g_k) \ \text{ s.t. the set } \{x \in \mathbb{R}^n : u(x) \geq 0\} \text{ is compact.} \quad (28)$$

This condition clearly implies that K is compact. Moreover, it does not depend on the set K but on the choice of the polynomials used to describe K. Note that it is easy to modify the presentation of K so that the condition (28) holds. Indeed, if we know the radius R of a ball containing K then, by adding to the description of K the (redundant) polynomial constraint $g_{k+1}(x) := R^2 - \sum_{i=1}^{n} x_i^2 \geq 0$, we can ensure that assumption (28) holds for this enriched presentation of K.

For convenience we recall the definition of the bounds $val_{outer}^{(r)}$ from (14):

$$val_{outer}^{(r)} = \inf_{\mu \in (Q^r(g_1, \ldots, g_k))^*} \left\{ \int_K f_0(x) d\mu(x) : \int_K f_i(x) d\mu(x) = b_i \ \forall i \in [m] \right\},$$

where we refer to (6) and (7) for the definitions of the truncated quadratic module $Q^r(g_1, \ldots, g_k)$ and of its dual cone $(Q^r(g_1, \ldots, g_k))^*$.

We also recall the stronger bounds $\overline{val}_{outer}^{(r)}$, introduced in (18), and obtained by replacing in the definition of $val_{outer}^{(r)}$ the cone $Q^r(g_1, \ldots, g_k)$ by the larger cone $Q^r(\prod_{j \in J} g_j : J \subseteq [k]))$, so that we have

$$val_{outer}^{(r)} \leq \overline{val}_{outer}^{(r)} \leq val.$$

5.1 Asymptotic and Finite Convergence

Here we present some results on the asymptotic and finite convergence of the lower bounds on val obtained by considering outer approximations of the cone $M(K)_+$.

Asymptotic Convergence

The parameters $val^{(r)}_{outer}$ form a non-decreasing sequence of lower bounds for the optimal value val of problem (1), which converge to it under assumption (28). This asymptotic convergence result relies on the following representation result of Putinar [45] for positive polynomials.

Theorem 8 (Putinar) *Assume K is compact and assumption (28) holds. Any polynomial f that is strictly positive on K (i.e., $f(x) > 0$ for all $x \in K$) belongs to $Q^r(g_1, \ldots, g_k)$ for some $r \in \mathbb{N}$.*

The following result can be found in [32, 33] for the general GPM and in [31] for the case of global polynomial optimization.

Asymptotic Convergence for the Bounds $val^{(r)}_{outer}$

Theorem 9 *Assume K is compact and assumption (28) holds. Then we have*

$$val^* \leq \lim_{r \to \infty} val^{(r)}_{outer} \leq val,$$

with equality: $val^* = \lim_{r \to \infty} val^{(r)}_{outer} = val$ if, in addition, there exists $z \in \mathbb{R}^{m+1}$ such that $\sum_{i=0}^{m} z_i f_i(x) > 0$ for all $x \in K$.

This result follows using Theorem 8. Observe that it suffices to show the inequality: $val^* \leq \sup_r val^{(r)}_{outer}$ (as the rest follows using Corollary 1). For this let $\epsilon > 0$ and let $y \in \mathbb{R}^m$ be feasible for val^*, i.e., $f_0(x) - \sum_{i=1}^{m} y_i f_i(x) \geq 0$ for all $x \in K$; we will show the inequality $b^T y \leq \sup_r val^{(r)}_{outer} + \epsilon\mu(K)$. Then, letting ϵ tend to 0 gives $b^T y \leq \sup_r val^{(r)}_{outer}$ and thus the desired result: $val^* \leq \sup_r val^{(r)}_{outer} = \lim_{r \to \infty} val^{(r)}_{outer}$.

As the polynomial $f_0 + \epsilon - \sum_i y_i f_i$ is strictly positive on K, it belongs to $Q^r(g_1, \ldots, g_k)$ for some $r \in \mathbb{N}$ in view of Theorem 8. Then, for any measure μ feasible for $val^{(r)}_{outer}$, we have $\int_K (f_0 + \epsilon - \sum_i y_i f_i)d\mu \geq 0$, which implies $b^T y \leq \int_K f_0 d\mu + \epsilon\mu(K)$ and thus the desired inequality:

$$b^T y \leq val^{(r)}_{outer} + \epsilon\mu(K) \leq \sup_r val^{(r)}_{outer} + \epsilon\mu(K).$$

When assuming only K compact (thus not assuming condition (28)), the following representation result of Schmüdgen [50] permits to show the asymptotic convergence of the stronger bounds $\overline{val}_{outer}^{(r)}$ to val (in the same way as Theorem 9 follows from Putinar's theorem).

Theorem 10 (Schmüdgen) *Assume K is compact. Any polynomial f that is strictly positive on K (i.e., $f(x) > 0$ for all $x \in K$) belongs to $Q^r (\prod_{j \in J} g_j : J \subseteq [k])$ for some $r \in \mathbb{N}$.*

Asymptotic Convergence for the Bounds $\overline{val}_{outer}^{(r)}$

Theorem 11 *Assume K is compact. Then we have*

$$val^* \leq lim_{r \to \infty} \overline{val}_{outer}^{(r)} \leq val,$$

with equality: $val^* = lim_{r \to \infty} \overline{val}_{outer}^{(r)} = val$ if, in addition, there exists $z \in \mathbb{R}^{m+1}$ such that $\sum_{i=0}^{m} z_i f_i(x) > 0$ for all $x \in K$.

Finite Convergence

A remarkable property of the lower bounds $val_{outer}^{(r)}$ is that they often exhibit finite convergence. Indeed, there is an easily checkable criterion, known as the *flatness condition*, that permits to conclude that the bound is exact: $val_{outer}^{(r)} = val$, and to extract an (atomic) optimal solution to the GPM. This is condition (29) below, which permits to claim that a given truncated sequence is indeed the sequence of moments of a positive measure; it goes back to work of Curto and Fialkow ([7], see also [33, 37] for details). To expose it we use the SDP formulation (16)–(17) for the parameter $val_{outer}^{(r)}$.

Finite Convergence

Theorem 12 (See [33, Theorem 4.1]) *Set $d_K := \max\{\lceil \deg(g_j/2 \rceil : j \in [k]\}$ and let $r \in \mathbb{N}$ such that $2r \geq \max\{\deg(f_i) : i \in \{0, \dots, m\}\}$ and $r \geq d_K$. Assume the program (16)-(17) defining the parameter $val_{outer}^{(r)}$ has an optimal solution $y = (y_\alpha)_{\alpha \in \mathbb{N}_{2r}^n}$ that satisfies the following (flatness) condition:*

$$rank M_s(y) = rank M_{s-d_K}(y) \quad \text{for some integer } s \text{ s.t. } d_K \leq s \leq r, \tag{29}$$

where

$$M_s(y) = (y_{\alpha+\beta})_{\alpha,\beta\in\mathbb{N}_s^n} \quad \text{and} \quad M_{s-d_K}(y) = (y_{\alpha+\beta})_{\alpha,\beta\in\mathbb{N}_{s-d_K}^n}.$$

Then equality $val_{outer}^{(r)} = val$ holds and the GPM problem (1) has an optimal solution $\mu \in M(K)_+$ which is atomic and supported on $rankM_s(y)$ points in K.

Under the flatness condition (29) there is an algorithmic procedure to find the atoms and weights of the optimal atomic measure (see, e.g., [33, 37] for details).

In addition, for the special case of the polynomial optimization problem (8), Nie [42] shows that the flatness condition is a generic property, so that finite convergence of the lower bounds $val_{outer}^{(r)}$ to the minimum of a polynomial over K holds generically.

Note that analogous results also hold for the stronger bounds $\overline{val}_{outer}^{(r)}$ on val.

5.2   Error Analysis for the Case of Polynomial Optimization

We now consider the special case of global polynomial optimization, i.e., problem (8), which is the case of GPM with only one affine constraint, requiring that μ is a probability measure on K:

$$val = \min_{x\in K} p(x) = \min_{\mu\in M(K)_+} \int_K p(x)d\mu(x) \text{ s.t. } \int_K d\mu(x) = 1.$$

Recall the definition of the bound $val_{outer}^{(r)}$ from (14), which now reads

$$val_{outer}^{(r)} = \inf_{\mu\in(Q^r(g_1,\ldots,g_k))^*} \left\{ \int_K p(x)d\mu(x) : \int_K d\mu(x) = 1 \right\}.$$

It can be reformulated via an SDP as in (16)–(17), whose dual SDP reads

$$\sup_{\lambda\in\mathbb{R}}\{\lambda : p - \lambda \in Q^r(g_1,\ldots,g_k)\}. \tag{30}$$

By weak duality $val_{outer}^{(r)}$ is at least the optimal value of (30). Strong duality holds for instance if the set K has a non-empty interior (since then the primal SDP is strictly feasible), or if there is a ball constraint present in the description of the set K (as shown in [27]). Then, $val_{outer}^{(r)}$ is also given by the program (30), which is the case, e.g., when K is a simplex, a hypercube, or a sphere.

As we saw above, the bounds $val_{outer}^{(r)}$ converge asymptotically to the minimum value val taken by the polynomial p over the set K when condition (28) holds. We now indicate some known results on the rate of convergence of these bounds.

For a polynomial $p = \sum_\alpha p_\alpha x^\alpha \in \mathbb{R}[x]_d$, we set

$$L_p := \max_\alpha |p_\alpha| \frac{\alpha_1! \cdots \alpha_n!}{|\alpha|!}.$$

Error Analysis for the Bounds $val_{outer}^{(r)}$

Theorem 13 ([43]) *Assume* $K \subseteq (-1, 1)^n$. *There exists a constant* $c > 0$ *(depending only on* K*) such that, for any polynomial* p *with degree* d, *we have*

$$val - val_{outer}^{(r)} \leq 6d^3 n^{2d} L_p \frac{1}{\left(\log \frac{r}{c}\right)^{1/c}} \quad \text{for all integers} \quad r \geq c \exp\left((2d^2 n^d)^c\right).$$

Note that this result displays a very slow convergence rate, which does not reflect the good behaviour of the bounds often observed in practice.

On the other hand, a sharper error analysis holds for the stronger bounds $\overline{val}_{outer}^{(r)}$, obtained by using the larger set $Q^r(\prod_{j \in J} g_j : J \subseteq [k])$ instead of $Q^r(g_1, \ldots, g_k)$.

Error Analysis for the Bounds $\overline{val}_{outer}^{(r)}$

Theorem 14 ([52]) *Assume* $K \subseteq (-1, 1)^n$. *There exists a constant* $c > 0$ *(depending only on* K*) such that, for any polynomial* p *with degree* d, *we have*

$$val - \overline{val}_{outer}^{(r)} \leq cd^4 n^{2d} L_p \frac{1}{r^{1/c}} \quad \text{for all integers} \quad r \geq cd^c n^{cd}.$$

We now recap some known sharper results for the case of polynomial optimization over special sets K like the simplex, the hypercube and the sphere. As motivation recall that this already captures well known hard combinatorial optimization problems such as the maximum independence number in a graph.

Given a graph $G = (V = [n], E)$ let $\alpha(G)$ denote the largest cardinality of an independent set in G, i.e., of a set $I \subseteq V$ that does not contain any edge of E. In fact the parameter $\alpha(G)$ can be reformulated via polynomial optimization over the simplex Δ_n, the hypercube $[0, 1]^n$, or the unit sphere S_n. Indeed the following results are known:

$$\frac{1}{\alpha(G)} = \min_{x \in \Delta_n} x^T (I_n + A_G)x, \quad \alpha(G) = \max_{x \in [0,1]^n} \sum_{i \in V} x_i - \sum_{\{i,j\} \in E} x_i x_j,$$

$$\frac{2\sqrt{2}}{3\sqrt{3}}\sqrt{1 - \frac{1}{\alpha(G)}} = \max_{y \in \mathbb{R}^n, z \in \mathbb{R}^m} \left\{ 2 \sum_{\{i,j\} \in \overline{E}} z_{ij} y_i y_j : (y, z) \in S_{n+m} \right\}$$

(see [40, 41]). Here I_n is the identity matrix of size n, A_G is the adjacency matrix of G (with entries $A_{ij} = A_{ji} = 1$ if $\{i, j\} \in E$ and 0 otherwise), $\overline{E}$ is the set of pairs of distinct elements $i, j \in V$ such that $\{i, j\} \notin E$ and $m = |\overline{E}|$.

Error Analysis for the Sphere

We first consider the case of the sphere $K = S_n = \{x \in \mathbb{R}^n : \sum_{i=1}^n x_i^2 = 1\}$. Then an error analysis for the bounds $val_{outer}^{(r)}$ is known when p is a homogeneous polynomial.

First, one may reduce to the case when p has even degree. Indeed, as shown in [21], if p has odd degree d then we have

$$\max \left\{ p(x) : \sum_{i=1}^n x_i^2 = 1 \right\} = \frac{d^{d/2}}{(d+1)^{(d+1)/2}} \max \left\{ x_{n+1} p(x) : \sum_{i=1}^{n+1} x_i^2 = 1 \right\}.$$

Another useful observation is that, for a homogeneous polynomial q of even degree d, q belongs to the truncated quadratic module of the sphere:

$$Q^r \left(\pm \left(1 - \sum_{i=1}^n x_i^2 \right) \right) = \Sigma_r + \left(1 - \sum_{i=1}^n x_i^2 \right) \mathbb{R}[x]$$

if and only if the polynomial $q(x) \left(\sum_{i=1}^n x_i^2 \right)^r$ is a sum of squares of polynomials (see [16]). Therefore, when p is a homogeneous polynomial of even degree $d = 2a$, the parameter $val_{outer}^{(r)}$ can be reformulated as

$$val_{outer}^{(r)} = \min \left\{ t : t \in \mathbb{R}, \ t \left(\sum_{i=1}^n x_i^2 \right)^r - \left(\sum_{i=1}^n x_i^2 \right)^{r-a} p(x) \in \Sigma_r \right\}. \tag{31}$$

Based on this, the following error bounds for the parameters $val_{outer}^{(r)}$ are shown in [21, 24] (for general polynomials) and in [17] (for even polynomials).

Theorem 15 *Let p be a homogeneous polynomial of even degree d.*

(i) ([21, 24]) There exist constants $C_{n,d}$ and $r_{n,d}$ (depending on n and d) such that

$$\min_{x \in S_n} p(x) - val_{outer}^{(r)} \leq \frac{C_{n,d}}{r} \quad \text{for all integers } r \geq r_{n,d}.$$

(ii) ([17]) If p is an even polynomial (i.e., of the form $p = \sum_{\alpha \in \mathbb{N}_{d/2}^n} p_\alpha x^{2\alpha}$), then the above holds where the constant $C_{n,d}$ depends only on d and $r_{n,d} = d$.

We briefly discuss the approach in [21], which in fact provides an error analysis for the larger range $val_{inner}^{(r)} - val_{outer}^{(r)}$.

For an integer a let $\mathrm{MSym}((\mathbb{R}^n)^{\otimes a})$ denote the set of matrices acting on $(\mathbb{R}^n)^{\otimes a}$ that are maximally symmetric, which means the associated $2a$-tensor is fully symmetric (i.e., invariant under the action of the symmetric group $\mathrm{Sym}(2a)$). Any homogeneous polynomial p of degree $2a$ can be written as $p(x) = (x^{\otimes a})^T Z_p x^{\otimes a}$ for a (unique) $Z_p \in \mathrm{MSym}((\mathbb{R}^n)^{\otimes a})$. Then, defining the polynomial $p_r(x) = (\sum_i x_i^2)^{r-a} p(x)$, the program (31) can be reformulated as

$$val_{outer}^{(r)} = \min\left\{ \langle Z_{p_r}, M \rangle : M \succeq 0, \mathrm{Tr}(M) = 1, \ M \in \mathrm{MSym}((\mathbb{R}^n)^{\otimes r}) \right\}.$$

Let M be an optimal solution to this program. As $M \succeq 0$ the polynomial $(x^{\otimes r})^T M x^{\otimes r}$ is a sum of squares. One can scale it to obtain $h \in \Sigma_r$ which provides a probability density function on S_n, i.e., $\int_{S_n} h(x) d\mu_0(x) = 1$ (with μ_0 the surface measure on S_n), and thus $val_{inner}^{(r)} \leq \int_{S_n} h(x) d\mu_0$. Using the orthogonal polynomial basis with respect to μ_0 (consisting of spherical harmonic polynomials), Doherty and Wehner [21] show a de Finetti type result, which permits to upper bound the range $\int_{S_n} h(x) d\mu_0 - \langle Z_{p_r}, M \rangle$ and thus $val_{inner}^{(r)} - val_{outer}^{(r)}$.

Error Analysis for the Simplex and the Hypercube
For the simplex $K = \Delta_n = \{x \in \mathbb{R}^n : x_i \geq 0 \ (i \in [n]), 1 - \sum_{i=1}^n x_i = 0\}$ and the hypercube $K = [0, 1]^n = \{x \in \mathbb{R}^n : x_i \geq 0, 1 - x_i \geq 0 \ (i \in [n])\}$, a refined error analysis is known only for the stronger bounds $\overline{val}_{outer}^{(r)}$, where we use the larger quadratic module generated by all pairwise products of the constraints defining K.

Error Analysis for the Simplex

Theorem 16 ([17]) *Assume $K = \Delta_n$ and p is a homogeneous polynomial with degree d. Then we have*

$$\min_{x \in \Delta_n} p(x) - \overline{val}_{outer}^{(r)} \leq \frac{C_d}{r} \left(\max_{x \in \Delta_n} p(x) - \min_{x \in \Delta_n} p(x) \right) \quad \text{for all } r \geq d,$$

where $C_d > 0$ is an absolute constant depending only on d.

Error Analysis for the Hypercube

Theorem 17 ([12]) *Assume $K = [0, 1]^n$. For any polynomial p with degree d we have*

$$\min_{x \in [0,1]^n} p(x) - \overline{val}_{outer}^{(r)} \leq n^d \binom{d+1}{3} L_p \frac{1}{r} \quad \text{for all } r \geq d.$$

The above results show that in Theorem 14 one may choose the unknown constant to be $c = 1$ (roughly) if K is a hypercube or simplex. In both cases the proof relies on showing this error analysis for a weaker bound, which is obtained by using only nonnegative scalar multipliers (instead of sum-of-squares multipliers) in the definition of the quadratic module. See [12, 17] for details.

6   Concluding Remarks

We conclude with a few remarks on available software and future research directions.

Software
The bounds based on the outer approximations (14) described here have been implemented in the software *Gloptipoly3* [26]. The software can in fact deal with a more general version of the GPM (1) than presented here. Namely it can deal with the problem

$$val = \inf_{\mu_i \in \mathcal{M}(K_i)_+ \, \forall i \in \{0\} \cup [m]} \left\{ \int_{K_0} f_0(x) d\mu_0(x) \; : \; \int_{K_i} f_i(x) d\mu_i(x) = b_i \quad \forall i \in [m] \right\},$$

where we have a variable measure $\mu_i \in \mathcal{M}(K_i)_+$ for each index $i \in \{0, \ldots, m\}$, with $K_i \subseteq \mathbb{R}^n$ being basic closed semi-algebraic sets defined by (possibly different) sets of polynomial inequalities.

Due to the sizes of the resulting semidefinite programs that are solved, applicability is typically limited to $n \leq 20$ variables and low order, say $r \leq 4$. This is due to the fact the matrix variables in the semidefinite programs are roughly of order $\binom{n+r}{r}$. Solving larger instances requires exploiting additional structure (like sparsity) leading to more economical semidefinite programs. We refer, e.g., to [33] and references therein for further details.

Error Bounds for the Inner Approximation Hierarchy
The known error bounds for the inner approximation, presented earlier in Table 2, are for specific choices of the set K and reference measure $\mu_0 \in \mathcal{M}(K)_+$. More work is required to understand the role of the reference measure in the convergence analysis, and to extend the regime in $O(1/r^2)$ to more classes of sets K. In particular, an obvious choice is whether one can sharpen the analysis of the the convergence rate for the Euclidean unit sphere. As explained, such results would also have implications for grid search on cubature points on the sphere. Cubature on the sphere is a vast research topic (see, e.g., [8, Chapter 6]), even in the special case of spherical t-designs [8, §6.5], where all cubature weights are equal and positive. Moreover, the complexity of polynomial optimization on spheres is not fully understood; indeed the problem is NP-hard, but allows polynomial-time approximation schemes in special cases (see [10, 17]). Sharpening the analysis of the inner approximations for polynomial optimization over spheres may help to gain a more complete understanding.

Error Bounds for the Outer Approximation Hierarchy
The bounds based on the outer approximation presented here are more practically suited for computation, in particular since they (sometimes) enjoy finite convergence and permit to extract the global minimizers; moreover, as mentioned above, the dedicated software *Gloptipoly3* is available for this purpose. On the other hand, the known results on the rate of convergence are somewhat disappointing (as discussed in Sect. 5.2), and in general much weaker than those known for the inner approximation. There is certainly room for a breakthrough here; new ideas are needed to obtain convergence rates that match the performance observed in practice.

Acknowledgements This work has been supported by European Union's Horizon 2020 research and innovation programme under the Marie Skłodowska-Curie grant agreement 813211 (POEMA).

The authors would like to thank Fernando Mario de Oliveira Filho for insightful discussions on the duality theory of the GPM.

Note Added in Proof Some of the above mentioned questions have been recently addressed. In particular, the results in Table 2 for the inner approximation bounds have been sharpened. Namely, the convergence rate in $O(1/r^2)$ has been extended for the sphere in [15] and for the hypercube equipped with more measures in [55]. A sharper rate in $O(\log^2 r/r^2)$ for convex bodies and in $O(\log r/r)$ for compact sets with an interior condition is shown in [55]. In addition, the convergence rate $O(1/r^2)$ is shown in [23] for the outer approximation bounds in the case of the unit sphere.

References

1. Akhiezer, N.I. *The Classical Moment Problem.* Hafner, New York (1965)
2. Barvinok, A. *A Course in Convexity.* Graduate Study in Mathematics, Volume 54, AMS, Providence, Rhode Island (2002)
3. Bayer, C., and Teichmann, J. The proof of Tchakaloff's theorem. *Proceedings of the American Mathematical Society*, 134:3035–3040 (2006)

4. Ben Tal, A., and Nemirovski, A. *Lectures on Modern Convex Optimization: Analysis, Algorithms, and Engineering Applications.* MPS/SIAM Series on Optimization, 2, SIAM (2001)
5. Castella, M. Rational optimization for nonlinear reconstruction with approximate ℓ_0 penalization. Preprint version available at arXiv:1808.00724 (2018)
6. Cools, R. An encyclopaedia of cubature formulas, *J. Complexity,* 19, 445–453 (2003).
7. Curto, R.E., and Fialkow, L.A. Solution of the truncated complex moment problem for flat data, *Memoirs of the American Mathematical Society* 119 (568) (1996)
8. Dai, F., and Xu, Y. *Approximation Theory and Harmonic Analysis on Spheres and Balls,* Springer, New York (2013)
9. De Klerk, E., Hess, R., and Laurent, M. Improved convergence rates for Lasserre-type hierarchies of upper bounds for box-constrained polynomial optimization. *SIAM Journal on Optimization* 27(1), 347–367 (2017)
10. De Klerk, E. The complexity of optimizing over a simplex, hypercube or sphere: a short survey. *Central European Journal of Operations Research,* 16(2), 111–125 (2008)
11. De Klerk, E., Lasserre, J.B., Laurent, M., and Sun, S. Bound-constrained polynomial optimization using only elementary calculations. *Mathematics of Operations Research,* 42(3), 834–853 (2017)
12. De Klerk, E., Laurent, M. Error bounds for some semidefinite programming approaches to polynomial minimization on the hypercube. *SIAM Journal on Optimization* 20(6), 3104–3120 (2010)
13. De Klerk, E., Laurent, M. Comparison of Lasserre's measure-based bounds for polynomial optimization to bounds obtained by simulated annealing. *Mathematics of Operations Research,* to appear. Preprint version available at http://arxiv.org/abs/1703.00744 (2017)
14. De Klerk, E., Laurent, M. Worst-case examples for Lasserre's measure–based hierarchy for polynomial optimization on the hypercube. *Mathematics of Operations Research,* to appear. Preprint version available at http://arxiv.org/abs/1804.05524 (2018)
15. de Klerk, E., Laurent, M. Convergence analysis of a Lasserre hierarchy of upper bounds for polynomial minimization on the sphere. arXiv:1904.08828 (2019)
16. De Klerk, E., Laurent, M., and Parrilo, P. On the equivalence of algebraic approaches to the miniization of forms on the simplex. In D. Henrion and A. Garulli (eds), *Positive Polynomials in Control,* 121–133, Springer (2005)
17. De Klerk, E., Laurent, M., and Parrilo, P. A PTAS for the minimization of polynomials of fixed degree over the simplex. *Theoretical Computer Science,* 361(2-3), 210–225 (2006)
18. De Klerk, E., Laurent, M., Sun, Z. Convergence analysis for Lasserre's measure-based hierarchy of upper bounds for polynomial optimization. *Mathematical Programming Series A* 162(1), 363–392 (2017)
19. De Klerk, E., Laurent, M., Sun, Z., and Vera, J. On the convergence rate of grid search for polynomial optimization over the simplex. *Optimization Letters,* 11(3), 597–608 (2017)
20. De Klerk, E., Postek, K., and Kuhn, D. Distributionally robust optimization with polynomial densities: theory, models and algorithms. Preprint version available at arXiv:1805.03588 (2018)
21. Doherty, A.C., Wehner, S. Convergence of SDP hierarchies for polynomial optimization on the hypersphere. arXiv:1210.5048v2 (2013)
22. Dunkl, C.F., and Xu., Y. *Orthogonal Polynomials of Several Variables,* Cambridge University Press (2001)
23. Fang, K., Fawzi, H. The sum-of-squares hierarchy on the sphere, and applications in quantum information theory. Preprint (2019)
24. Faybusovich, L. Global optimization of homogeneous polynomials on the simplex and on the sphere. In C. Floudas and P. Pardalos (eds), *Frontiers in Global Optimization,* Kluwer (2003)
25. Folland, G.B. How to integrate a polynomial over a sphere? *The American Mathematical Monthly,* 108(5), 446–448 (2001)

26. Henrion, D., Lasserre, J.B., and Loefberg, J. GloptiPoly 3: moments, optimization and semidefinite programming. *Optimization Methods and Software*, 24(4-5), 761–779 (2009). Software download: www.laas.fr/$\sim$henrion/software/gloptipoly3

27. Josz, C., Henrion, D. Strong duality in Lasserre's hierarchy for polynomial optimization. *Optim. Letters*, 10, 3–10 (2016)

28. Kalai. A. T., and Vempala, S. Simulated annealing for convex optimization. *Mathematics of Operations Research*, 31(2), 253–266 (2006)

29. Kemperman, J.H.B. The general moment problem, a geometric approach. *The Annals of Mathematics Statistics*, 39, 93–122 (1968)

30. Landau, H. *Moments in Mathematics, Proc. Sympos. Appl. Math.*, 37 (1987)

31. Lasserre, J.B. Global optimization with polynomials and the problem of moments. *SIAM J. Optim.* 11, 796–817 (2001)

32. Lasserre, J.B. A semidefinite programming approach to the generalized problem of moments. *Mathematical Programming Series B* 112, 65–92 (2008)

33. Lasserre, J.B. *Moments, Positive Polynomials and Their Applications.* Imperial College Press (2009)

34. Lasserre, J.B. *Introduction to Polynomial and Semi-Algebraic Optimization.* Cambridge University Press (2015)

35. Lasserre, J.B. A new look at nonnegativity on closed sets and polynomial optimization. *SIAM Journal on Optimization* 21(3), 864–885 (2011)

36. Lasserre, J.B. The moment-SOS hierarchy. *Proc. Int. Cong. of Math. ? 2018*, Rio de Janeiro, 3, 3761–3784 (2018)

37. Laurent, M. Sums of squares, moment matrices and optimization over polynomials. In *Emerging Applications of Algebraic Geometry,* Vol. 149 of IMA Volumes in Mathematics and its Applications, M. Putinar and S. Sullivant (eds.), Springer, 157–270 (2009)

38. Laurent, M. Optimization over polynomials: Selected topics. In Chapter 16 (Control Theory and Optimization) of *Proc. Int. Cong. of Math. 2014*. Jang, S. Y., Kim, Y. R., Lee, D-W. & Yie, I. (eds.). Seoul: Kyung Moon SA Co. Ltd., p. 843–869 (2014)

39. Martinez, A., Piazzon, F., Sommariva, A., and Vianello, M. Quadrature-based polynomial optimization. Optim. Lett. (2019). https://doi.org/10.1007/s11590-019-01416-x

40. Motzkin, T.S., Sraus, E.G. Maxima for graphs and a new proof of a theorem of Túran. *Canadian J. Math.*, 17, 533–540 (1965)

41. Nesterov, Yu. Random walk in a simplex and quadratic optimization over convex polytopes. CORE Discussion Paper 2003/71, CORE-UCL, Louvain-La-Neuve (2003)

42. Nie, J. Optimality conditions and finite convergence of Lasserre's hierarchy. *Mathematical Programming, Ser. A,* 146(1-2), 97–121 (2014)

43. Nie, J., and Schweighofer, M. On the complexity of Putinar's positivstellensatz *Journal of Complexity* 23, 135–150 (2007)

44. Piazzon, F., Vianello, M. Markov inequalities, Dubiner distance, norming meshes and polynomial optimization on convex bodies. Preprint at Optimization Online (2018)

45. Putinar, M. Positive polynomials on compact semi-algebraic sets. *Ind. Univ. Math. J.* 42, 969–984 (1993)

46. Putinar, M. A note on Tchakaloff's theorem. Proceedings of the American Mathematical Society, 125(8), 2409–2414 (1997)

47. Reznick, B. Some concrete aspects of Hilbert's 17th Problem. In *Real algebraic geometry and ordered structures (Baton Rouge, LA, 1996)*, pages 251–272. Amer. Math. Soc., Providence, RI, 2000.

48. Rogosinski, W.W. Moments of non-negative mass, *Proceedings of the Royal Society A* 245, 1–27 (1958)

49. Ryu, E.K. and Boyd, S.P. Extensions of Gauss Quadrature Via Linear Programming. *Foundations of Computational Mathematics* 15(4), 953–971 (2015)

50. Schmüdgen, K. The K-moment problem for compact semi-algebraic sets. *Math. Ann.*, 289, 203–206 (1991)

51. Schmüdgen, K. *The Moment Problem*. Springer (2017)
52. Schweighofer, M. On the complexity of Schmüdgen's Positivstellensatz, *Journal of Complexity* 20(4), 529–543 (2004)
53. Schwartz, R.E. The 5 electron case of Thomson's problem. *Exp. Math* 22(2), 157–186 (2013)
54. Shapiro, A. On duality theory of conic linear problems, *Semi-Infinite Programming: Recent Advances (M,Á. Goberna and M.A. López, eds.)*, Springer, 135–165 (2001)
55. Slot, L., Laurent, M. Improved convergence analysis of Lasserre's measure-based upper bounds for polynomial minimization on compact sets. arXiv:1905.08142 (2019)
56. Tao, T. An Epsilon of Room, I: Real Analysis: pages from year three of a mathematical blog. AMS, Graduate Studies in Mathematics Volume: 117 (2010)
57. Trefethen, L.N. Cubature, approximation, and isotropy in the hypercube. *SIAM Review*, 59(3), 469–491 (2017)
58. Tchakaloff, V. Formules de cubature mécanique à coefficients non négatifs, *Bull. Sci. Math.*, 81, 123–134 (1957)

Algebra and Geometry in the Study
of Enzymatic Cascades

Alicia Dickenstein

Abstract In recent years, techniques from computational and real algebraic geometry have been successfully used to address mathematical challenges in systems
biology. The algebraic theory of chemical reaction systems aims to understand their
dynamic behavior by taking advantage of the inherent algebraic structure in the
kinetic equations, and does not need a priori determination of the parameters, which
can be theoretically or practically impossible. This chapter gives a brief introduction
to general results based on the network structure. In particular, we describe a
general framework for biological systems, called MESSI systems, that describe
Modifications of type Enzyme-Substrate or Swap with Intermediates and include
many post-translational modification networks. We also outline recent methods to
address the important question of multistationarity, in particular in the study of
enzymatic cascades, and we point out some of the mathematical questions that arise
from this application.

1 Introduction

We start by introducing the cartoon mechanisms of two enzymatic signalign
pathways depicted in research articles.The important RAS signaling pathway in
Fig. 1 includes an extracellular ligand and a transmembrane receptor, which trigger
a cascade of protein-protein interactions and enzymatic reactions, then integrated
into key biological responses controlling cell proliferation, differentiation or death.
When this pathway is altered, it can drive to unhealthy cell proliferation [41].
Figure 2 presents a more precise description of the last part of the enzymatic
cascade.

A. Dickenstein (✉)
Department of Mathematics, FCEN, University of Buenos Aires, Ciudad Universitaria, Pab. I,
Buenos Aires, Argentina

IMAS, UBA-CONICET, Ciudad Universitaria, Pab. I, Buenos Aires, Argentina
e-mail: alidick@dm.uba.ar

© The Association for Women in Mathematics and the Author(s) 2019
C. Araujo et al. (eds.), *World Women in Mathematics 2018*, Association for Women
in Mathematics Series 20, https://doi.org/10.1007/978-3-030-21170-7_2

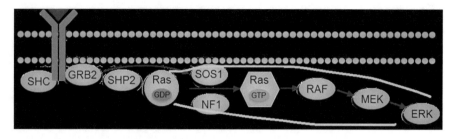

Fig. 1 The RAS signaling pathway, starting in the membrane of the cell

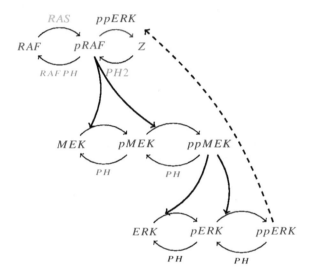

Fig. 2 Part of the RAS signaling pathway inside the cell, possibly with retroactivity

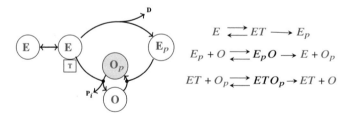

Fig. 3 EnvZ-OmpR bacterial model

Figure 3 depicts an osmolarity regulation network in bacteria, which is implemented in part by the EnvZ/OmpR two-component system [49]. The sensor kinase EnvZ (denoted by E in the diagram) autophosphorylates on a histidine residue (E_p) and catalyzes the transfer of the phosphate group to the aspartate residue of the response regulator OmpR (O), which then acts as an effector. In this mechanism, when EnvZ is bounded to ATP (ET), it also catalyzes hydrolysis of the phosphorylated OmpR-P (O_p), which is a transcription factor that regulates the expression of various protein pores. This unusual design keeps the limit concentration of OmpR-P at a value that is independent of the positive initial concentrations.

When we first look at these biological mechanisms, it does not seem evident that algebra and geometry can be used to analyze them. But we will argue in this chapter that this is indeed the case and that we can contribute with these mathematical tools to the understanding of questions in Systems Biology.

In particular, in the realm of biochemical reaction networks, that is, chemical reaction networks in biochemistry, the usual mass-action kinetics modeling of the evolution of the concentrations of the different chemical species along time (as RAS, RAF, MEK, ERK, E, O, etc. above) yields an autonomous system of polynomial ordinary differential equations $\frac{dx}{dt} = f_\kappa(x)$ in the unknown vector of concentrations x of the species as functions of time, for each choice of the (real positive) reaction rate constants κ (see Definition 1). In fact, these equations are associated to a labeled directed graph G of reactions. The monomial terms come from the labels of the nodes of G by complexes in the given species, the coefficients depend on the (positive) reaction rate constants κ that label the edges of G, and the total production of each reaction (which is the difference of the labels of the target and source nodes). The real polynomials $f_\kappa(x)$ carry a combinatorial structure inherited from G and we will also think of κ as *parameters* and consider the *family* of differential systems parametrized by them. Chemical Reaction Network Theory (CNRT) was initiated by Horn and Jackson and subsequently by Feinberg and his students and collaborators [22] and has seen a great development over the last years, when new combinatorial and algebro-geometric techniques have been introduced. We refer the reader to the survey article [16] for basic definitions, results and further references, and we review here some advances developed after that article was published.

In Sect. 4 we recall the notion of MESSI systems we introduced in [42]. Many post-translational modification networks are MESSI networks. For example: the motifs in [23], sequential distributive multisite networks [52], sequential processive multisite phosphorylation networks [12], phosphorylation cascades or the bacterial EnvZ/OmpR network from [49] in Fig. 3. Our work is inspired by and extends some results in several previous articles [24, 28, 29, 31, 39, 43, 48, 51]. MESSI is an acronym for Modifications of type Enzyme-Substrate or Swap with Intermediates (see Definition 2). Networks with an underlying MESSI structure include many post-translational modification networks, as well as all linear systems arising from mass-action kinetics (a.k.a. Laplacian dynamics [38]). We summarize some results and algorithms based on this structure to predict conservation relations, persistence, the capacity for multistationarity, and the description of regions of multistationarity. Once the network has the capacity for multistationarity, the next main question is how to predict parameters of, if possible, regions in parameter space which give rise to multistationary systems, which are called *multistationarity regions*. In Sect. 5 we comment on several recent approaches to study multistationarity in chemical reaction networks. Section 6 mentions the mostly unexplored question of the a priori determination of the occurrence of oscillations in chemical reaction networks, in particular, in enzymatic networks. We end the paper with two main open questions.

2   Basics of Mass-Action Kinetics

In this section we set the basic terminology and the mathematical concepts mentioned in the introduction. In particular, we discuss the notion of multistationarity.

Two-component signal transduction systems enable bacteria to sense, respond, and adapt to a wide range of environments, stressors, and growth conditions. Before giving the precise Definition 1, we instantiate mass-action kinetics in a biological example of a simple two-component mechanism. It relies on *phosphotransfer* reactions. Upon receiving a signal, the *hybrid histidine kinase* HK can self-phosphorylate. This is a hybrid histidine kinase with two phosphorylatable domains. We denote the phosphorylation state of each site by p, if the site is phosphorylated, and 0, if it is not; the four possible forms are HK_{00}, HK_{p0}, HK_{0p}, HK_{pp}. The *response regulator protein* is denoted by RR when it is unphosphorylated and RR_p denotes the phosphorylated form. Given a vector of *reaction rate constants* $k = (k_1, \ldots, k_6) \in \mathbb{R}^6_{>0}$, the (directed) graph of reaction equals:

$$HK_{00} \xrightarrow{k_1} HK_{p0} \xrightarrow{k_2} HK_{0p} \xrightarrow{k_3} HK_{pp}$$

$$HK_{0p} + RR \xrightarrow{k_4} HK_{00} + RR_p$$

$$HK_{pp} + RR \xrightarrow{k_5} HK_{p0} + RR_p$$

$$RR_p \xrightarrow{k_6} RR,$$

where each of the ten nodes corresponds to a *complex* on the six *chemical species*, that we number in the following order: HK_{00}, HK_{p0}, HK_{0p}, HK_{pp}, RR, RR_p. Mass-action kinetics specifies how the respective *concentrations* $x_1, \ldots, x_6$ of these six species evolve with time. The basic principle in this modeling is derived from the idea that the rate of an elementary reaction is proportional to the probability of collision of the reactants, which under an independence assumption equals the product of their concentrations. We derive the following autonomous polynomial dynamical system $\frac{dx_i}{dt} = f_i(x)$, $i = 1, \ldots, 6$:

$$\frac{dx_1}{dt} = -k_1 x_1 + k_4 x_3 x_5, \qquad \frac{dx_2}{dt} = k_1 x_1 - k_2 x_2 + k_5 x_4 x_5,$$

$$\frac{dx_3}{dt} = k_2 x_2 - k_3 x_3 - k_4 x_3 x_5, \qquad \frac{dx_4}{dt} = k_3 x_3 - k_5 x_4 x_5,$$

$$\frac{dx_5}{dt} = -k_4 x_3 x_5 - k_5 x_4 x_5 + k_6 x_6, \qquad \frac{dx_6}{dt} = k_4 x_3 x_5 + k_5 x_4 x_5 - k_6 x_6.$$

It is straightforward to check that the following linear dependencies hold and generate all the linear dependencies among $f_1, \ldots, f_6$:

$$f_1 + f_2 + f_3 + f_4 = f_5 + f_6 = 0,$$

from which we deduce two linear *conservation relations*:

$$x_1 + x_2 + x_3 + x_4 = T_1, \qquad x_5 + x_6 = T_2.$$

Thus, trajectories lie in a 4-plane in 6-space. The *total conservation constants* T_1, T_2 are determined by the initial conditions $(x_1(0), \ldots x_6(0))$.

Given a numbering of the species as above, we usually identify a complex on these species with a nonnegative integer vector. For example, the complex $y = X_3 + X_5$ is identified with the vector $e_3 + e_5 = (0, 0, 1, 0, 1, 0) \in \mathbb{Z}^6_{\geq 0}$. The general definition is as follows.

Definition of Chemical Reaction Networks and Mass-Action Kinetics

Definition 1 A *chemical reaction network* (on a finite set of s species, which we assume ordered) is a finite labeled directed graph $G = (V, E, (\kappa_{ij})_{(i,j) \in E}, (y_i)_{i=1,\ldots,m})$, whose vertices V are labeled by complexes $y_1, \ldots, y_m \in \mathbb{Z}^s_{\geq 0}$ and whose edges $(i, j) \in E$ are labeled by positive real numbers $i \overset{\kappa_{ij}}{\to} j$. We will also say that G is a network.

Mass-action kinetics specified by the network G gives the following autonomous system of ordinary differential equations in the *concentrations* $x = (x_1, x_2, \ldots, x_s)$ of the species as functions of time:

$$\frac{dx}{dt} = \sum_{(i,j) \in E} \kappa_{ij} \, x^{y_i} \, (y_j - y_i) = f_\kappa(x). \tag{1}$$

Here, $\frac{dx}{dt}$ and $y_j - y_i$ are column vectors.

Note that the coordinates $f_1, \ldots, f_s$ of f_κ are polynomials in $\mathbb{R}[x_1, \ldots, x_s]$ (to ease the notation we omit the dependence of f_i on κ). Many systems occurring in population dynamics, for example the oscillatory Lotka-Volterra equations, can be viewed as arising from a chemical reaction network as in (1), but for instance not the "chaotic" Lorenz equations. A simple characterization of autonomous dynamical systems arising from chemical reaction networks under mass-action kinetics has been given by Hárs and Tóth. We refer to the book [20], which also contains an introduction to the stochastic modeling of chemical kinetics.

Another direct consequence of the form of the equations in (1) is that for any trajectory $x(t)$, the vector $\frac{dx}{dt}$ lies for all t (in any interval I containing 0 where it is defined) in the so called *stoichiometric subspace* S, which is the linear subspace generated by the differences $\{y_j - y_i \mid (i, j) \in E\}$. Using the shape of the polynomials f_i it can be seen that the positive orthant $\mathbb{R}^s_{>0}$ and its closure $\mathbb{R}^s_{\geq 0}$ are forward-invariant for the dynamics. Then, any trajectory $x(t)$ starting at a nonnegative point $x(0)$ lies for all $t \in I \cap \mathbb{R}_{>0}$ in the closed polyhedron

$(x(0) + S) \cap \mathbb{R}_{\geq 0}^s$, which is called a *stoichiometric compatibility class*, or for short, an *S-class*.

Denote by q the codimension of S. Given a basis $\ell_1, \ldots, \ell_q$ of linear forms in the dual of S, let $T_i = \ell_i(x(0))$, $i = 0, \ldots, q$. The equations $\ell_1(x) = T_1, \ldots, \ell_q(x) = T_q$ of $x(0) + S = S_T$ give linear *conservation relations* and, as above, the constant coefficient T_i of such a linear equation is called a *total conservation constant*.

The Steady State Variety and the Notion of Multistationarity

The *steady state variety* $V_\kappa(f)$ of the kinetic system (1) equals the nonnegative real zeros of $f_1, \ldots, f_s$:

$$V_\kappa(f) = \{x \in \mathbb{R}_{\geq 0}^s : f_1(x) = \cdots = f_s(x) = 0\}. \tag{2}$$

An element of $V_\kappa(f)$ is called a *steady state* of the system and corresponds to a constant trajectory in the nonnegative orthant. We say that system (1) exhibits *multistationarity* if there exist at least two *positive* steady states with the same total conservation constants, that is, in the same S-class. This is an important property for chemical reaction networks modeling biological processes, since the ocurrence of multistationarity allows for different *responses* of the cell under the same total conservation constants, depending on the initial conditions.

In fact, our point of view will be the following. The underlying reaction network $(V, E, (y_i)_{i=1,\ldots,m})$ defines a *family* of autonomous polynomial dynamical systems depending on the positive parameters $\kappa \in \mathbb{R}_{>0}^{\#E}$. We say that it has the *capacity for multistationarity* if there is a choice of reaction rate constants $\kappa = (\kappa_{ij})_{(i,j)\in E}$ and total conservation constants $T = (T_1, \ldots, T_q)$ for which the intersection of the steady state variety $V_\kappa(f)$ with the *positive* points of linear variety S_T consists of more than one point (that is: there exist parameters κ and T such that there are at least two points in the positive orthant lying in the intersection of the steady state variety $V_\kappa(f)$ with the S-class defined by T).

There are many results to decide the capacity for multistationarity of a given chemical reaction network, starting with [14]. Most of them have been summarized in Theorem 1.4 of [39]. In fact, these results give in general necessary and sufficient conditions for the stronger condition that the map f_κ is injective on the positive points of all S-classes. There are several implementations of different algorithms, starting with the pioneering algorithm implemented by Feinberg and his group in the Chemical Reaction Network Toolbox. The link to the corresponding webpage together with links to other algorithms can be found at https://reaction-networks.net/wiki/Mathematics_of_Reaction_Networks#. We recall some of the tools to address this question in Sects. 4 and 5.

In Fig. 4, there is a range of values of T for which there are three positive steady states on the corresponding translate S_T of S (i.e., in an S-class) for a fixed value

Fig. 4 The green curve represents the steady state variety $V_{\kappa^*}(f)$. The subspace $S = \{\ell = 0\}$ is a line. The number of points of intersection of the translates $S_T = \{\ell = T\}$ of S with $V_{\kappa^*}(f)$ in the positive orthant depends on the total conservation constant T

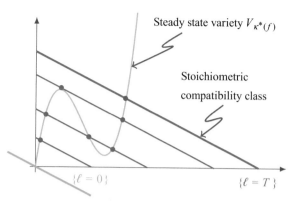

Steady state variety $V_{\kappa^*}(f)$

Stoichiometric compatibility class

$\{\ell = 0\}$ $\{\ell = T\}$

Fig. 5 Only one parameter is allowed to vary

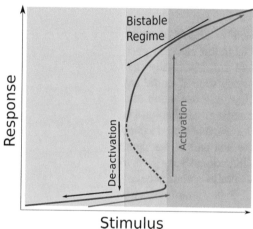

Bistable Regime

Response

De-activation

Activation

Stimulus

κ^* of positive rate constants. So, the chemical reaction network has the capacity for multistationarity and κ^* is a choice of *multistationarity parameter*.

We feature two kinds of multistationarity pictures from the literature. One way to find the special values rendering these figures is by measurements in experiments or by exhaustive (and lucky) simulations of the trajectories taking sample values in the space of parameters and initial conditions. Instead, one can try to develop algebro-geometric tools to analyze the mathematical models arising from biochemical reaction networks, with the goal of making predictions from the structure of the networks.

Figure 5 corresponds to a 2-site sequential phosphorylation and dephosphorylation that we describe in Sect. 3 below. This network has 15 parameters: 12 reaction constants and 3 total conservation constants. In the picture, all the reaction rate constants and two of the total conservation constants have been specialized and only the total conservation constant E_{tot} of one enzyme is varying. This is considered to be the *input* variable (or *stimulus*) and it is represented on the x-axis. The number of chemical species is equal to 9, but only one of the phosphorylated substrates s^* at

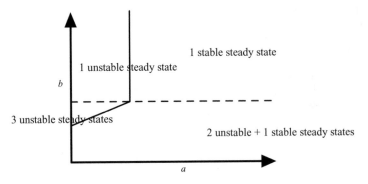

Fig. 6 Only two of the parameters are allowed to vary

steady state is represented, which is consider the *response* of the system. It happens that in this case any positive value of s^* is one coordinate of a positive steady state and different steady states in the same S-class have different s^* coordinates. The steady state s^*-coordinate is represented on the y-axis. For small or big values of E_{tot}, only one value of s^* is possible, so this is a monostationary regime. In the middle zone, there are three steady states, two *stable* and one unstable, so this is the bistable regime (stability of steady states is determined by the negativity of the real part of the eigenvalues of the Jacobian). This figure corresponds to a two dimensional very particular "slice" of points originally in $24 = 15 + 9$ variables, where 14 variables have been specialized and 8 variables are not shown.

Figure 6 represents a two dimensional "slice", but in *parameter space*, of another mechanism that we do not specify, but in which only two of the parameters (a, b) are allowed to vary. For each of the values of (a, b) outside the line segments separating the regions, there are either one or three positive steady states, which could be stable or unstable. In fact, in most biochemical networks these curves separating the regions are *far from being line segments*; they are high order algebraic hypersurfaces that separate different semialgebraic regions where the qualitative dynamics is the same, in a high dimensional parameter space. Moreover, regions with interesting behaviour could be small.

The separating hypersurfaces related to the question of multistationarity are described by the union of the discriminant associated to the equations describing $V_\kappa(f)$ and S_T with respect to the x variables (which vanishes whenever there is a point where the intersection of the steady state variety and the S-class is non-transversal), and the union for any $i \in \{1, \ldots, s\}$ of the resultant describing the fact that there is a common point with $x_i = 0$. In each chamber (connected component) of the complement of the union of these algebraic varieties, the number of real roots is the same and moreover, for each of the real roots it holds that the sign of each of the coordinates does not change as the parameters are moved, and thus the number of real roots with a fixed sign (for instance, positive roots) is constant along the chamber. We refer the reader to the book [26] for the notions of discriminant and resultant, which are in general not linear. These polynomials in the parameters can be computed effectively—in theory—via different computational algebraic geometry methods of elimination of variables, but standard computations are not feasible when there are many variables. Even if one can compute these equations, it is a very complicated task to describe then all the possible chambers in the complement of its zero locus, or at least to find one representative in each chamber. There are implementations by M. Safey El Din, which work very well in small examples using his package RAGlib [47].

3 Two Important Families of Enzymatic Networks

In this section, we introduce common enzymatic mechanisms that will help us exemplify and clarify the concepts we will introduce in Sect. 4.

Sequential Phosphorylations

The multisite n-phosphorylation system describes the site phosphorylation of a protein (with n sites where a phosphate group can be absorbed or emitted) by a pair of enzymes (a kinase and a phosphatase) in a sequential and distributive mechanism. The Nobel Prize in Physiology or Medicine was awarded in 1992 to Edmond Fischer and Edwin Krebs "for their discoveries concerning reversible protein phosphorylation as a biological regulatory mechanism." The kinase and the phosphatase speed up the transformation of other proteins without being incorporated in the final products of the process, which is crucial in the regulation of metabolism in the body. Multi-site phosphorylation plays important regulatory roles in cell cycle regulation and inflammation pathways, and is implicated in multiple disorders, including Alzheimer disease. Because of the important role played by these systems in signal transduction networks inside the cell, there is a body of work on the mathematics of phosphorylation systems (which belong to the more general class of post-translational modification systems). We refer the reader to the papers [33, 43, 50] and the references therein.

We now describe the special case of a sequential phosphorylation/dephosphory-lation with $n = 2$ sites, which is also known as the dual futile cycle. There are nine species: three substrates (the unphosphorylated substrate S_0, the substrate with one and two phosphorylated sites S_1 and S_2), two enzymes (the kinase E and the phosphatase F), and four intermediate species (ES_0, ES_1, FS_2 and FS_1). We give to the twelve rate constants the usual names in the literature [52].

$$S_0 + E \underset{k_{off_0}}{\overset{k_{on_0}}{\rightleftarrows}} ES_0 \overset{k_{cat_0}}{\longrightarrow} S_1 + E \underset{k_{off_1}}{\overset{k_{on_1}}{\rightleftarrows}} ES_1 \overset{k_{cat_1}}{\longrightarrow} S_2 + E$$

$$S_2 + F \underset{l_{off_1}}{\overset{l_{on_1}}{\rightleftarrows}} FS_2 \overset{l_{cat_1}}{\longrightarrow} S_1 + F \underset{l_{off_0}}{\overset{l_{on_0}}{\rightleftarrows}} FS_1 \overset{l_{cat_0}}{\longrightarrow} S_0 + F$$

We number the species and their concentrations as follows: x_1, x_2, x_3 denote the respective concentrations of S_0, S_1, S_2; y_1, y_2, y_3, y_4 denote the respective concentrations of the intermediate species ES_0, ES_1, FS_2, FS_1, x_4 is the concentration of the kinase E, and x_5 the concentration of the phosphatase F. The associated system of ODE's defined in (1) equals in this case:

$$\frac{dx_1}{dt} = -k_{on_0} x_1 x_4 + k_{off_0} y_1 + l_{cat_0} y_4$$

$$\frac{dx_2}{dt} = -k_{on_1} x_2 x_4 + k_{cat_0} y_1 + k_{off_1} y_2$$

$$- l_{on_0} x_2 x_5 + l_{cat_1} y_3 + l_{off_0} y_4$$

$$\frac{dx_3}{dt} = k_{cat_1} y_2 - l_{on_1} x_3 x_5 + l_{off_1} y_3$$

$$\frac{dy_1}{dt} = k_{on_0} x_1 x_4 - (k_{off_0} + k_{cat_0}) y_1$$

$$\frac{dy_2}{dt} = k_{on_1} x_2 x_4 - (k_{off_1} + k_{cat_1}) y_2$$

$$\frac{dx_4}{dt} = -k_{on_0} x_1 x_4 - k_{on_1} x_2 x_4 + (k_{off_0} + k_{cat_0}) y_1$$

$$+ (k_{off_1} + k_{cat_1}) y_2$$

$$\frac{dx_5}{dt} = -l_{on_0} x_2 x_5 - l_{on_1} x_3 x_5 + (l_{off_1} + l_{cat_1}) y_3$$

$$+ (l_{off_0} + l_{cat_0}) y_4$$

$$\frac{dy_3}{dt} = l_{on_1} x_3 x_5 - (l_{off_1} + l_{cat_1}) y_3$$

$$\frac{dy_4}{dt} = l_{on_0} x_2 x_5 - (l_{off_0} + l_{cat_0}) y_4.$$

There are 3 independent linear conservation laws, for instance:

$$x_1 + x_2 + x_3 + y_1 + y_2 + y_3 + y_4 = S_{tot}$$

$$x_4 + y_1 + y_2 = E_{tot}$$

$$x_5 + y_3 + y_4 = F_{tot},$$

where $S_{tot}, E_{tot}, F_{tot}$ are positive real numbers for any choice of initial condition in the positive orthant. As we pointed out in Sect. 1, there are $12 + 3 = 15$ parameters. The n-site sequential mechanism is similar, with $3n + 3$ variables, $6n$ reaction rate constants and always 3 total conservation constants, so a total of $6n + 3$ parameters.

Phosphorylation Cascades

We have already encountered a coarse diagram of an enzymatic cascade in Fig. 2. MAP kinase cascades are important signal transduction systems in molecular biology for which there is also a body of mathematical work, see for instance [35, 41] and the references therein. These cascades correspond to a network of enzymatic reactions arranged in layers, where usually in each of them there is a futile cycle of sequential phosphorylations and such that the fully phosphorylated substrate serves as an enzyme for the next layer.

The simplest case of a cascade with the capacity of multistationarity [23] consists of a cascade with two layers and a single phosphorylation/dephosphorylation at each layer, with one phosphatase. It corresponds to the a labeled digraph, with 9 variables and 18 parameters, where each single phosphorylation follows the same mechanism as in our previous example, with an intermediate species. The nine species are the substrates S_0, S_1 in the first layer, the substrates P_0, P_1 in the second layer, four intermediate complexes, a kinase E and the *same* phosphatase F to dephosphorylate the substrates in both layers. The forward enzyme in the second layer is the phosphorylated substrate S_1 from the first layer.

$$S_0 + E \underset{k_{\text{off}_0}^1}{\overset{k_{\text{on}_0}^1}{\rightleftarrows}} E S_0 \overset{k_{\text{cat}_0}^1}{\longrightarrow} S_1 + E$$

$$S_1 + F \underset{l_{\text{off}_0}^1}{\overset{l_{\text{on}_0}^1}{\rightleftarrows}} F S_1 \overset{l_{\text{cat}_0}^1}{\longrightarrow} S_0 + F$$

$$P_0 + S_1 \underset{k_{\text{off}_0}^2}{\overset{k_{\text{on}_0}^2}{\rightleftarrows}} S_1 P_0 \overset{k_{\text{cat}_0}^2}{\longrightarrow} P_1 + S_1$$

$$P_1 + F \underset{l_{\text{off}_0}^2}{\overset{l_{\text{on}_0}^2}{\rightleftarrows}} F P_1 \overset{l_{\text{cat}_0}^2}{\longrightarrow} P_0 + F.$$

This mechanism is usually depicted as follows, hiding the reaction rate constants and the intermediate species:

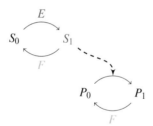

In this case, there are 4 linearly independent conservation relations. Denoting with small letters the concentration of each of the species, these conservation relations can be chosen as follows, as predicted in Theorem 3.2 in [42] (see (4) below):

$$s_0 + s_1 + e s_0 + f s_1 + s_1 p_0 = S_{\text{tot}}$$
$$p_0 + p_1 + s_1 p_0 + f p_1 = P_{\text{tot}}$$
$$e + e s_0 = E_{\text{tot}}$$
$$f + f s_1 + f p_1 = F_{\text{tot}},$$

where $S_{tot}, P_{tot}E_{tot}, F_{tot}$ are positive real numbers for any choice of initial condition in the positive orthant.

We can also consider cascades with any number n of layers. In this case, the number of variables, the number of reaction rate constants and the number of independent linear conservation relations (as well as the number of linear conservation constants) grow linearly with n.

4 MESSI Systems

In this section we recall the notion of MESSI networks from [42], to describe a common structure underlying the four examples above in their different variants as well as many "popular" biological networks, that consist of Modifications of type Enzyme-Substrate or Swap with Intermediates. The occurrence of this structure allows us to prove general results for quite different mechanisms. The basic ingredient of a MESSI structure is a partition of the set of species, which reflects the different chemical behaviors. This grouping of the chemical species into disjoint subsets is in accordance with the intuitive partition of the species according to their function that biochemists have. We will denote the disjoint union of sets with the symbol $\bigsqcup$.

Definition of a MESSI System

Definition 2 A MESSI network is a chemical reaction network satisfying the following properties. First of all, there exists a partition of the set S of species

$$S = S^{(0)} \bigsqcup S^{(1)} \bigsqcup S^{(2)} \bigsqcup \cdots \bigsqcup S^{(m)}, \tag{3}$$

where $m \geq 1$, $S^{(0)}$ is the subset of intermediate species and could be empty, and all $S^{(i)}$ with $i \geq 1$ are nonempty subsets, formed by what we call core species. We requiere that the complexes and reactions satisfy the following conditions. An intermediate species can only be part of a monomolecular complex consisting only of this speces (called an intermediate complex). Non-intermediate complexes are called core complexes and consist of one or otherwise two chemical species belonging to different subsets of the partition. Denote by $y \to_\circ y'$ the existence of an edge from complex y to complex y' or a directed path of reactions from y to y' through intermediate complexes. We require that for any intermediate complex y_0, there exist core complexes y, y' such that $y \to_\circ y_0 \to_\circ y'$. If there are two monomolecular core complexes $y \to_\circ y'$, then both should consist of a species in the same $S^{(\alpha)}$. We further ask that if there is a reaction between a monomolecular and a bimolecular complex, the monomolecular complex is an intermediate, and that if y, y' are bimolecular core complexes such that $y \rightarrow_\circ y'$, then there exist two different core subsets $S^{(\alpha)}$, $S^{(\beta)}$ in the partition, such that both y and y' consist of a species in each of them.

When endowed with mass-action kinetics, a MESSI network gives rise to a MESSI system of polynomial autonomous ODE's.

All the Networks We Mentioned Are MESSI

All the networks we mentioned in the text (plus many other common biochemical networks) can be endowed with the structure of a MESSI system. We gave different colors to the different subsets in a possible partition of the species.

For instance, in the cascade depicted in Fig. 2 in the Introduction, the intermediate species (complexes) are not displayed, but we presented with different colors a possible partition of the core species that defines a MESSI structure. In the network depicted in Fig. 3 the partition into a subset of intermediate species (in black), and two subsets of core species (in red and blue) also defines a MESSI structure.

In the two-component system in Sect. 2, we could take $S^{(0)} = \emptyset$, $S^{(1)} = \{HK_{00}, HK_{p0}, HK_{0p}, HK_{pp}\}$, and $S^{(2)} = \{RR, RR_p\}$.

In the example of the sequential phosphorilation in Sect. 3, we could take $S^{(0)} = \{ES_0, ES_1, FS_2, FS_1\}$, $S^{(1)} = \{S_0, S_1, S_2\}$; $S^{(2)} = \{E\}$, and $S^{(3)} = F$. It can be checked that all conditions are satified. Note that if we consider the coarser partition with the same set of intermediate species $S^{(0)}$, the same set $S^{(1)}$ of core species, and just one other set $\{E, F\}$ of core species, we also have a MESSI structure. In fact, there is in general a *poset* of possible partitions (and in other examples there could be non-comparable partitions).

On the other side, in the example of the cascade in Sect. 3, we can partition the set of nine species as follows to define a MESSI structure in the 2-layer cascade: $S^{(0)}$ consists of the four intermediate species $\{ES_0, FS_1, S_1P_0, FP_1\}$, plus the core subsets $S^{(1)} = \{S_0, S_1\}$, $S^{(2)} = \{P_0, P_1\}$, $S^{(3)} = \{E\}$, and $S^{(4)} = \{F\}$.

Conservation Laws

The first general results about MESSI systems is that we can describe enough (explicit) conservation linear relations with positive coefficients. Given a partition (3) of the set S of variables into one intermediate subset and $m \geq 1$ nonempty core subsets defining a MESSI structure in a given network G, note that the associated autonomous polynomial dynamical system defined in (1) is linear in the variables of each $S^{(i)}$ union the subset Int_i consisting of those intermediate species y' for which there exists a core complex y containing one species of $S^{(i)}$ such that $y \rightarrow_\circ y'$ (for any fixed $i = 1, \ldots, m$). The union of these subsets Int_i equals $S^{(0)}$, but they are in general not disjoint, because if in the recent notation y also contains a species in another $S^{(j)}$, then y' also belongs to Int_j. These intersections account for several important properties of the systems.

Theorem 3.2 in [42] asserts that given a partition of $S = \{x_1, \ldots, x_s\}$ defining a MESSI structure as in (3), the following linear forms $\ell_1, \ldots, \ell_m$ belong to the dual of the stoichiometric subspace S:

$$\ell_i(x) = \sum_{x_j \in S^{(i)}} x_j + \sum_{x_j \in \text{Int}_i} x_j, \quad i = 1, \ldots, m. \tag{4}$$

We refer the reader to Section 3 in [42] for conditions ensuring that these are a basis of conservation relations (and examples where this is not the case). We conclude that all MESSI systems are *conservative*. Thus, all S-classes are compact, and all trajectories are bounded and defined for any positive time. In fact, given a MESSI network, if x is a trajectory of the associated mass-action kinetics dynamical system $\dot{x}(t) = f(x(t))$, for all t in an open interval containing $\mathbb{R}_{\geq 0}$) with $x(0) \in \mathbb{R}_{>0}^s$, let $(T_1, \ldots, T_m) = (\ell_1(x(0)), \ldots, \ell_m(x(0)))$. Then, we have that for any $t \geq 0$ it holds that $\ell_i(x(t)) = T_i$ for any i. Then, all the coefficients of the linear form $\ell = \sum_{i=1}^{m} \ell_i$ are positive and $\ell(x(t)) = \sum_{i=1}^{m} T_i > 0$.

The Associated Digraphs

In order to state some other general results for MESSI networks, we introduce three associated digraphs G_1, G_2, G_E associated with a given MESSI network G with a vector of rate constants k. We refer the reader to Section 3 in [42] for complete definitions, explanations and examples.

We eliminate all intermediate species to define G_1, which naturally inherits a MESSI structure: the species of G_1 are the core species of G, its complexes are the core complexes of G and there is an edge between two core complexes y, y' precisely when $y \to_\circ y'$ in G. The rate constants of G_1 are rational functions $\tau(\kappa)$ with nonzero denominator over all positive κ, in such a way that when viewed with mass action kinetics gives rise to a system of the form $x' = f^1(x')$, the steady state variety $V_\tau(\kappa)(f^1)$ of the system defined by G_1 is a projection of the steady state variety $V_\kappa(f)$ of the original system. They have been explicitly defined in display (15) of the Supplementary Material in [24], see displays (5.3) and (5.8) in [6]. To define the digraph G_2, we first consider for any $i = 1, \ldots, m$ the linear network obtained by "hiding" in the rate constants the concentration of all species $x_j \notin S^{(i)}$. For instance, an edge $X_j + X_k \to X_{j_1} + X_{k_1}$ with $X_j, X_{j_1} \in S^{(i_1)}$, $X_k, X_{k_1} \in S^{(i_2)}$, with rate constant c, gives raise to the following two edges in G_2: the edge $X_j \to X_{j_1}$ with rate constant cx_j, and the edge $X_k \to X_{k_1}$, with rate constant cx_k. Note that we get this way a multidigraph MG_2 with possibly repeated edges and loops. We then denote by G_2 the digraph derived from MG_2 after collapsing multiple edges into a single edge, with label equal to the sum of the labels of the different edges. The nodes in each connected component of G_2

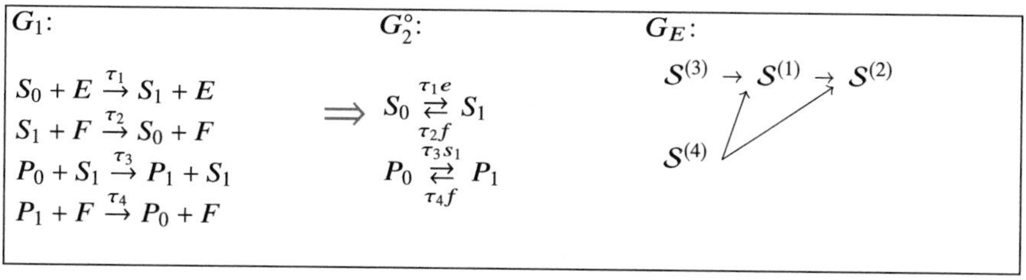

G_1:

$$S_0 + E \xrightarrow{\tau_1} S_1 + E$$
$$S_1 + F \xrightarrow{\tau_2} S_0 + F$$
$$P_0 + S_1 \xrightarrow{\tau_3} P_1 + S_1$$
$$P_1 + F \xrightarrow{\tau_4} P_0 + F$$

$\Longrightarrow$

G_2°:

$$S_0 \underset{\tau_2 f}{\overset{\tau_1 e}{\rightleftarrows}} S_1$$
$$P_0 \underset{\tau_4 f}{\overset{\tau_3 s_1}{\rightleftarrows}} P_1$$

G_E:

$$S^{(3)} \to S^{(1)} \to S^{(2)}$$
$$S^{(4)} \nearrow$$

Fig. 7 The graphs G_1, G_2° and G_E for the phosphorylation cascade in Sect. 3

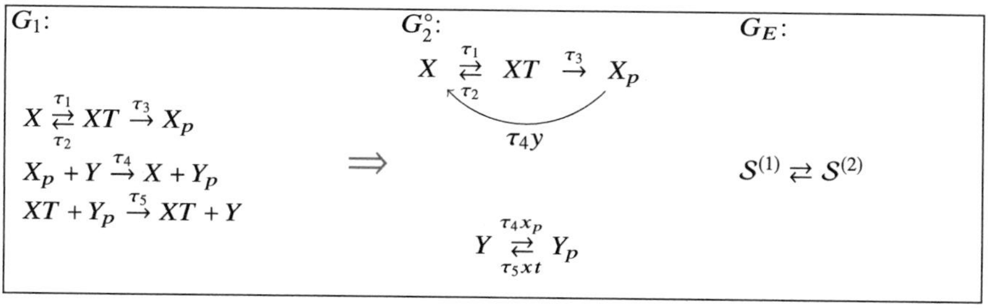

G_1:

$$X \underset{\tau_2}{\overset{\tau_1}{\rightleftarrows}} XT \xrightarrow{\tau_3} X_p$$
$$X_p + Y \xrightarrow{\tau_4} X + Y_p$$
$$XT + Y_p \xrightarrow{\tau_5} XT + Y$$

$\Longrightarrow$

G_2°:

$$X \underset{\tau_2}{\overset{\tau_1}{\rightleftarrows}} XT \xrightarrow{\tau_3} X_p$$
$$\tau_4 y$$
$$Y \underset{\tau_5 xt}{\overset{\tau_4 x_p}{\rightleftarrows}} Y_p$$

G_E:

$$S^{(1)} \rightleftarrows S^{(2)}$$

Fig. 8 The graphs G_1, G_2° and G_E for the EnvZ/OmpR two-component network in the Introduction

correspond to the species in one of the subsets $S^{(i)}$ of the partition if and only if this partition is minimal (in the poset of partitions of S defining a MESSI structure on G). The digraph G_2 is linear (each node is labeled with a monomolecular complex with a single species) and again, if we formally associate to it mass-action kinetics, its steady state variety coincides with that of G_1. Finally, we denote by G_2° the multidigraph obtained from G_2 after deleting all loops. On the other side, the nodes of the digraph G_E are the subsets $S^{(1)}, \ldots, S^{(m)}$ and there is an edge from $S^{(i_1)}$ to $S^{(i_2)}$ with a label containing as a factor the concentration of any species in $S^{(i_1)}$.

The graphs G_1, G_2° and G_E associated to two of the networks in the previous sections are depicted in Figs. 7 and 8.

Persistence

A chemical reaction system (1) is persistent if any trajectory starting from a point with positive coordinates stays at a positive distance from any point in the boundary, or informally, if no species which is present can tend to be eliminated in the course of the reaction. A steady state lying in the boundary of the nonnegative orthant (that is, with some coordinates equal to zero) is called *relevant* if it lies in the intersection of the boundary of the nonnegative orthant with a stoichiometric compatibility class through a point in $\mathbb{R}^s_{>0}$. As MESSI systems are conservative, Theorem 2 in [1] proves that a MESSI system is persistent when there are no *relevant boundary steady states*.

Given a MESSI network G, we identify the following hypotheses:

(A) The associated digraph G_2 is weakly reversible.
(B) The associated digraph G_E has no directed cycles.

Hypothesis (A) means that for any pair of nodes in the same connected component, there is a directed path from one to the other. For instance, in the two examples considered in Figs. 7 and 8, hypothesis (A) is verified. Hypothesis (B) is also verified in the case of the cascade network, but not in the EnvZ/OmpR two-component network. However, even if they sound restrictive, there is a big range of signaling pathways that satisfy both hypotheses.

Theorem 3.15 in [42] asserts that a MESSI network G which satisfies hypotheses (A) and (B) does not have relevant boundary steady states, and is thus persistent. Moreover, as MESSI systems are conservative, a version of Brouwer's fixed point theorem ensures the existence of a non-negative steady state in each S-class. So, the abscence of relevant boundary steady states implies the existence of a positive steady state in each S-class.

Explicit Parametrization of $V_\kappa(f) \cap \mathbb{R}^s_{>0}$

We describe a big class of MESSI networks for which the steady state variety V is rational. This is a very uncommon property for general algebraic varieties.

Explicit Rational Parametrizations

We want to describe the intersection $V_\kappa(f) \cap S_T$ in the positive orthant. The steady state variety is defined in principle by s polynomial equations. Assume the dimension of S (and thus of S_T for any T) equals $s - q$ and can thus be defined by q linear equations. This implies that there are (at most) $s - q$ linearly independent polynomials among $f_1, \ldots, f_s$. A finite number of common solutions is expected, but this might not be true.

One way to simplify the computation of the intersection is the following. As S_T are linear varieties, they can be parametrized by $s - q$ parameters. One could then parametrize S_T solving for q variables in terms of the other ones and then replace this in the equations of the steady state variety. This reduces the number of variables from s to $s - q$, but the polynomials $f_1, \ldots, f_s$ are particular, with a monomial structure that comes from G and we would in general destroy the sparsity.

Denote by $V_{>0,\kappa}(f) = V_\kappa(f) \cap \mathbb{R}^s_{>0}$. One could then try to parametrize $V_{>0,\kappa}(f)$ but general algebraic varieties do not have rational parametrizations. This is a very uncommon property for general algebraic varieties. However, rational parametrizations do exist for the positive points of the steady state variety in certain enzymatic biochemical networks, as proved by Thomson and Gunawardena in [51]. We extended this result for many other networks of biological interest which are MESSI. Theorem 4.1 in [42] proves the existence of an explicit and algorithmically constructible rational parametrization of $V_{>0,\kappa}(f)$ for any MESSI network G satisfying conditions (A) and (B) above. Moreover, if the partition is minimal with m subsets of core species, we have that $\dim V_{>0,\kappa}(f) = m = s - \dim S$.

Moreover, we identify conditions that ensure that this parametrization is monomial, or equivalently, that $V_{>0,\kappa}(f)$ can be cut out by binomial equations (that is, polynomials with two terms) and, in this case, we give explicit binomials in Theorem 4.8 in [42] for what we call s-toric MESSI systems. Again, the conditions seem to be very restrictive, but there are plenty of interesting signaling pathways that satisfy them; for instance the n-site phosphorilation networks and many enzymatic cascades, as the ones we presented in Sect. 3. In the case of the n-sequential phosphorylation network (which has $3n + 3$ variables) we can parametrize the positive steady state variety with 3 parameters for any value of n. To compute the intersection $V_\kappa(f) \cap S_T$ (which equals $V_{>0,\kappa}(f) \cap S_T$ due to the abscence of relevant boundary steady states, as we pointed out before), we can write 3 of the variables in terms of the remaining $3n$ variables from the 3 conservation relations and replace them into $3n$ linearly independent f_i (which exist in this case). We could substitute the parametrization into the conservation relations and thus get 3 equations in 3 variables. This is what makes the n-site amenable to computations even if in principle the number or variables tends to infinity with n. Note that if instead we plug in a parametrization of S_T into the equations of the steady state variety, we get a system, that besides losing sparsity, consists of $3n$ equations in $3n$ variables.

Recognizing the existence of a MESSI structure on a given network, checking the hypotheses in all our results and finding the rational parametrization are algorithmic and only depend on the structure and not on the particular parameters.

Deciding Multistationarity

The important biological mechanism of n *sequential* phospho-dephosphorylations has the capacity for multistationarity for $n = 2$, that is, there can be up to 3 positive steady states in $V_\kappa(f) \cap S_T$ (for particular choices of the rate constants κ and positive linear conservation constants T). This system has been first studied by L. Wang and E. Sontag in [52]. They proved that the maximal possible number of positive steady states is $2n - 1$ and identified parameters for which there are $n + 1$ positive steady states for n even (and n for n odd). Note that $n + 1 = 2n - 1$ for $n = 2$. It has been proved in [36] that the upper bound $2n - 1$ is attained for $n = 3, 4$, and it is probable that $2n - 1$ is a sharp upper bound, but this has not been proven yet for $n \geq 5$. See also [33–35] for a discussion of other dynamical features (stability and oscillations).

In fact, the steady states of most popular MESSI systems (including all those recalled above) present an s-toric structure, and we gave in this case a characterization of the capacity for multistationarity, which lead to an algorithm based on tools from oriented matroid theory. The main ideas in this approach, which go back to [14] and several other papers, including articles in other applied areas, are collected and clarified in the paper [39]. We give below a simple version of the multistationarity results in Section 5 in [42], which is valid for other biochemical reaction networks for which the positive steady states can be defined by binomials in a parametric way and satisfying certain conditions (that we can ensure from the structure of the network, see e.g. Proposition 5.6 in [42]). In particular, these binomials are of the form $p_\kappa = a(\kappa)x^\alpha - b(\kappa)x^\beta$, with $\alpha, \beta \in \mathbb{Z}_{\geq 0}^s$, and a, b polynomial functions on the vector of rate constants $\kappa \in \mathbb{R}_{>0}^r$ taking positive values over $\mathbb{R}_{>0}^r$.

Given such a binomial p_κ, consider the vector $v_{p_\kappa} = \alpha - \beta \in \mathbb{Z}^s$ (note that $v_{p_\kappa} = -v_{-p_\kappa}$, so indeed v_{p_κ} are integer vectors defined up to sign). Also, given a matrix M of size $m_1 \times m_2$ of rank m_1, a subset J of indices of cardinality m_1 determines a maximal minor of M, which we denote by M_J.

Deciding Mono/Multistationarity

Let G be a chemical reaction network. Denote by $S^\perp$ a matrix whose rows define the dual of the stoichiometric subspace S with $\mathrm{rank}(S^\perp) = d$. Assume that $V_{>0,\kappa}(f)$ is cut out by $s - d$ binomials $p_{j,\kappa}$, $j = 1, \ldots, s - d$, with exponents $v_{p_{j,\kappa}}$ which form the columns of a matrix B. Assume moreover that $\mathrm{rank}(B) = s - d$. Then, the following statements are equivalent

1. Monostationarity: There is at most a single positive solution in $V_{>0,\kappa}(F) \cap S_T$, for any S-class intersecting the positive orthant, for *any* $\kappa \in \mathbb{R}_{>0}^r$.
2. For all subsets $J \subseteq \{1, \ldots, s\}$ of cardinality d, the product

$$(-1)^{\sum_{j\in J} j} \det(S_J^\perp) \det(B_{\{1,\ldots,s\}\setminus J})$$

either is zero or has the same sign as all other nonzero products, and at least one such product is nonzero.

The previous result can be turned into an algorithm to decide if a network has the capacity for multistationarity, together with an algorithm to produce vectors of rate constants k for which multistationarity occurs (in case the network is not monostationary).

5   Other Approaches to the Question of Multistationarity

The reader might have noticed that within a reasonable extension for a survey, we cannot properly define and explain all concepts. This section will then be only a pointer to some recent papers addressing the question of multistationarity, besides the articles and tools we have mentioned before. We also refer the reader to the recent survey [13] and the references therein.

Craciun, Helton and Williams applied in [15] the homotopy invariance of degree to determine the number of equilibria of biochemical reaction networks and how this number depends on parameters in the model. Conradi, Feliu, Mincheva and Wiuf give in [8] *necessary and sufficient conditions* for the multistationarity of networks having a *positive rational parametrization*, in terms of the reaction rate constants, also based on degree theory. This approach is very interesting since they can describe open multistationarity regions in rate constant space. However, it does not describe particular stoichiometric compatibility classes for which there is multistationarity, as it is also the case with the methods based on signs as the result we described about mono/multistationarity. The reason is that all these approaches are related (in more explicit or hidden ways) to properties of a Jacobian, for instance of an appropriate choice of the polynomials $f_1, \ldots, f_s$ and linear functions $\ell_1 - T_1, \ldots, \ell_q - T_q$ giving equations for S_T with respect to the x variables, and so the linear conservation constants $T_1, \ldots, T_q$ do not appear. In [18] we considered extensions and simplifications of this approach via critical functions, for networks with special structure, in particular for special MESSI networks which are commonly used in modeling enzymatic pathways. We also propose a method based on the existence of triangular forms, relying on techniques from computational algebra.

Sadeghimanesh and Feliu provide in [46] a new determinant criterion to decide whether a network is multistationary, when the network obtained by removing intermediates has a binomial steady state ideal. In this case, they characterize the *multistationarity structure of the network*, i.e. which subsets of complexes are responsible for multistationarity. In particular, they compute the multistationarity structure of the n-site sequential distributive phosphorylation cycle for any n.

Together with Bihan and Giaroli, we incorporated in [6] a new tool from real algebraic geometry based on the article [7] by Bihan, Santos, and Spaenlehauer. The basic idea is the following. Given a sparse polynomial system, that is, with exponents in a specified finite set of integer points A, if it is possible to find p *decorated* simplices in a *regular* subdivision of A, then it is possible to scale the coefficients of the given system in an explicit way to get *at least* p nondegenerate positive real roots. This gives a lower bound on the number of positive roots. The hypotheses of regularity of the subdivision means that it comes from a lifting of the points in A after considering the projection of the domains of linearity of the lower convex hull of the lifted points. This is what gives the necessary compatibility to find a common open set in the space of coefficients where the p positive solutions can be jointly continued. The meaning that a simplex is decorated is the following. Let $\{a_0, \ldots, a_d\} \subset A$ denote the set of vertices of a maximal dimensional simplex in dimension d. Given (Laurent) polynomials $g_1, \ldots, g_d$ with support A, consider their subsums of monomials corresponding only to these exponents. So one gets a system with d polynomials in d variables and $d + 1$ monomials of the form:

$$\sum_{j=0}^{d} c_j^i\, x^{a_i} = 0,\ i = 1,\ldots,d.$$

This system has at most one positive root and it does have a (nondegenerate) positive root exactly when the following linear system does:

$$c_0^i + \sum_{j=1}^{d} c_j^i\, x_i = 0,\ i = 1,\ldots,d.$$

This condition is equivalent to an alternance of signs of the minors of the $d \times (d+1)$ real matrix with coefficients c_{ij}. The simplex is said to be decorated by a choice of coefficients of the input polynomials when this is the case. It is interesting to note that, differently from the case of complex roots with nonzero coordinates, it is not always true that the lower bound in the case of positive solutions matches the maximum number of positive real roots for any regular subdivision. A simple example is the following. Assume $A = \{(0, 0), (1, 0), (1, 2), (2, 1)\}$ are the vertices of a paralellogram of Euclidean volume 2 in the plane. A sparse polynomial system $(g_1 = g_2 = 0)$ with this support can have $2 \cdot 2 = 4$ isolated complex solutions with nonzero coordinates by Kouchnirenko's theorem and 3 positive solutions (and this number can be attained, see [5] and the references therein). But it is clear that the support can only have three regular subdivisions: either nothing is subdivided or we get any of the two subdivisions depicted in Fig. 9, so the maximum lower bound p that one can obtain is 2. Nevertheless, this is up to now the only systematic way to find conditions on jointly on all the parameters that ensure the existence of several positive steady states, as for instance degree considerations are eventually based on parity considerations. But the best advantage of this approach is that it allows us to describe *multistationarity regions* in the space of all parameters, both reaction rate constants and linear conservation constants. Remark however that our conditions are only sufficient.

We refer the reader to Section 3 in [27] for a simple example explaining the technical results in [6]. These tools allowed us to find in that article precise multistationarity regions in enzyme cascades with any number n of layers of Goldbeter-Koshland loops (with a single phosphorylation/dephosphorylation in each layer), which are multistationary as soon as the two first phosphatases are the same. Interestingly, the number of variables is of the order of $4n$ and the dimension

Fig. 9 The two proper subdivisions of a circuit

of the stoichiometric subspace S is of the order of $2n$, so it is cut out by roughly $2n$ linear equations and parametrized by a similar number of variables. So, even taking advantage of the parametrizations of the steady state variety and a translate S_T of S, we need to deal with of the order of $2n$ equations in $2n$ variables. When the two layers with the same phosphatase are the last ones, it is possible to find particular multstationarity reaction rate constants for the cascade following the approach in [4]. Other papers based on the study of extrapolation of multistationarity from that of simpler subnetworks are for instance [9, 37].

In ongoing work with Giaroli, Pérez Millán and Rickster [17], we are able to use this setting to give a precise region in the space of all parameters for which the n-sequential phospho/dephosphorylation mechanism can have $n + 1$ for n even (and n for n odd) positive steady states, assuming that only $\frac{1}{4}$ of the intermediate complexes are part of the reactions. In another recent work Conradi, Iosif, and Kahle [10] also use tools from polyhedral geometry. They show that for reaction networks whose positive steady states can be cut out by binomials, multistationarity is scale invariant in the space of linear conservation constants (that is, if there is multistationarity for some value of the linear concentration constants, then there is multistationarity on the entire ray containing this value (possibly for different reaction rate constants). They consider the chamber decomposition in linear conservation constant space, which allows them to show that for values of these constants in one of the five chambers the 2-site sequential phosphorylation network cannot be multistationary.

Other approaches use numeric or symbolic methods to detect points in different chambers of the complement of the discriminant and the resultants that we mentioned before, see for instance [30, 32]. The general mathematical problem is the search of positive roots of sparse polynomial systems; see for instance [21] where these techniques have been applied to a geometric problem.

Stability and Convergence

The important question of deciding stability of a given steady state x^* of a chemical reaction network with fixed constants k^* can be formalized via Routh-Hurwitz theorem by means of the satisfiability of certain polynomial inequalities which correspond to minors of the Jacobian matrix at the point x^*, as a pattern of signs of these minors corresponds to all eigenvalues of the Jacobian having negative real part. However, this is a difficult question if the point x^* is given implicitly and if one tries to trace these inequalities as the parameters vary. So, only in few cases there is a complete analysis (see for instance [33]).

Another important question is to ensure convergence of the trajectories. Note that if a trajectory defined on the whole positive real line converges for $t \to +\infty$ to a point p, then p is a steady state. A first question is to decide global convergence in the presence of a single steady state in each S-class. We refer the reader to the results (and the references) in [19] for diverse architectures of processive multisite phoshorylation networks, which are based on previous work by Angeli, De Leenher and Sontag [2].

6   Oscillations

Another important biological feature is the possible occurrence of oscillations. Oscillations have been observed experimentally in signaling networks formed by phosphorylation and dephosphorylation, which seems to be the main mechanism in the 24-hour period in eukaryotic circadian clocks (see for instance [11, 44] and the references therein). Despite the many articles studying sequential phospho/dephosphorylation networks, it is not currently known whether in the 2-site sequential mechanism there could be trajectories which oscillate.

Instead, Suwanmajo and Krishnan showed recently in [50] that oscillations occur intrinsically in the the dual-site phosphorylation and dephosphorylation network, in which the mechanism for phosphorylation is processive while the one for dephosphorylation is distributive (or vice-versa), arising from a Hopf bifurcation. We also refer to the interesting paper [45], where the authors propose a systematic analysis of the long-term dynamics of phosphorylations systems. They describe bistability and oscillations when the network has nonzero levels of reaction processivity. Processivity means that the intermediate complex does not dissociate into substrate plus enzyme after a phospho/dephosphorylation, but only after two or more. Conradi, Mincheva, and Shiu showed in [11] for the mixed mechanism in [50] that in the three-dimensional space of linear conservation constants, the border between the existence of a stable or an unstable steady state is defined by the vanishing of a single Hurwitz determinant, which consists generically of simple Hopf bifurcations. Besides the Routh-Hurwitz criterion, their analysis relies on an algebraic Hopf-bifurcation criterion due to Yang and a monomial parametrization of the positive steady state variety. It would be very interesting to extend these kind of analyses to other mechanisms, in particular, to other phosphorylation networks.

Rendall and Hell studied in [34, 35] the existence of parameters for which Hopf bifurcations occur and generate periodic orbits in the case of (MAP kinase) cascades. They also explain how geometric singular perturbation theory allows to generalize results from simple models to more complex ones. Also Banaji presents in [3] some results are presented on how oscillation is inherited by chemical reaction networks (CRNs) when they are built in natural ways from smaller oscillatory networks, showing a particularly nice result for fully open networks (where for any species X, there are reactions $0 \to X$ and $X \to 0$), also based on regular and singular perturbation theory. We also mention the pioneering work of Karin Gatermann introducing algebraic and combinatorial techniques for the search of Hopf bifurcations [25].

7 Mathematical Challenges

In this section we enumerate some of the main open questions in this area. They involve difficult mathematical questions and moreover, systems of biological interest usually have a big number of variables and parameters.

1. Give general precise bounds for the number of positive solutions of (parametric families of) sparse polynomial systems and apply them to find the number of positive steady states: (a) develop tools to obtain better *lower bounds* for the number of positive steady states; (b) develop tools to get good *upper bounds* for the number of positive steady states. Moreover, find regions in parameter space with the predicted number of positive steady states, or at least where lower/upper bounds apply.
2. Predict or preclude oscillations from structure: how do (sustained) oscillations arise in phosphorylation networks? Can we find "atoms of oscillation"? Moreover, describe "regions of oscillation" in parameter space.

Conclusion

We can use algebro-geometric notions and methods to analyze system biology models. Algebraic and combinatorial methods allow us to predict (some) qualitative dynamic behaviours of our models from the structure of the network, without simulations and without measuring all the parameters a priori. We do have several promising results, but in many cases they tend to be too complex to be understood or computed. Answers to the above questions would require to develop a combination of tools from dynamical systems, real algebraic geometry, computational and numerical algebraic geometry, differential algebra, and biochemistry!

Acknowledgements I am very grateful to the Committee for Women in Mathematics of the International Mathematical Union for the organization of this meeting and for the many activities they take in charge to ensure development and recognition of female mathematicians around the world. I am particulary grateful to Carolina Araujo for all her work to make (WM)^2 a great success. I also thank Magalí Giaroli, Mercedes Pérez Millán and Enrique Tobis for their help with the figures.

References

1. Angeli D., De Leenher P., and Sontag. E: A Petri net approach to the study of persistence in chemical reaction networks, Mathematical Biosciences 210, 598–618 (2007).
2. Angeli D., De Leenher, P., and Sontag, E.: Graph-theoretic characterizations of monotonicity of chemical networks in reaction coordinates. J. Math. Biol. 61, 581–616 (2010).
3. Banaji, M.: Inheritance of oscillation in chemical reaction networks. Applied Mathematics and Computation 325, 191–209 (2018).

4. Banaji, M., and Pantea, C.: The inheritance of nondegenerate multistationarity in chemical reaction networks. SIAM J Appl Math 78(2), 1105–1130 (2018).
5. Bihan F., and Dickenstein, A.: Descartes' Rule of Signs for Polynomial Systems supported on Circuits, Int. Math. Res. Notices 22, 6867–6893 (2017).
6. Bihan F., Dickenstein, A., and Giaroli, M.: Lower bounds for positive roots and regions of multistationarity in chemical reaction networks. arXiv:1807.05157 (2018).
7. Bihan, F., Santos, F., and Spaenlehauer, P-J.: A polyhedral method for sparse systems with many positive solutions. SIAM J. Appl. Algebra Geometry 2(4), 620–645 (2018).
8. Conradi, C., Feliu, E., Mincheva, M., and Wiuf, C.: Identifying parameter regions for multistationarity. PLoS Comput. Biol. 13(10), e1005751 (2017).
9. Conradi, C., Flockerzi, D., Raisch, J., and Stelling, J.: Subnetwork analysis reveals dynamic features of complex (bio)chemical networks PNAS 104 (49), 19175–19180 (2007).
10. Conradi, C., Iosif, A., Kahle, T.: Multistationarity in the space of total concentrations for systems that admit a monomial parametrization. To appear: Bull. Math. Biol. (2019).
11. Conradi, C., Mincheva, M., Shiu, A: Emergence of oscillations in a mixed-mechanism phosphorylation system. To appear: Bull. Math. Bio. (2019).
12. Conradi, C., and Shiu, A.: A global convergence result for processive multisite phosphorylation systems. Bull. Math. Biol. 77(1), 126–155 (2015).
13. Conradi C., and Pantea C.: Multistationarity in Biochemical Networks: Results, Analysis, and Examples. Algebraic and Combinatorial Computational Biology, Ch. 9, Eds. Robeva R. and Macaulay M., Mathematics in Science and Computation, Academic Press (2019).
14. Craciun, G., and Feinberg. M., Multiple equilibria in complex chemical reaction networks. I. The injectivity property. SIAM J. Appl. Math. 65(5), 1526–1546 (2005).
15. Craciun, G., Helton, J. W., and Williams, R.: Homotopy methods for counting reaction network equilibria. Math. Biosci. 216(2), 140–149 (2008).
16. Dickenstein, A.: Biochemical reaction networks: an invitation for algebraic geometers. MCA 2013, Contemporary Mathematics 656, 65–83 (2016).
17. Dickenstein, A., Giaroli, M., Pérez Millán, M., and Rischter, R.: Parameter regions that give rise to $2 \left[\frac{n}{2} \right] + 1$ positive steady states in the n-site phosphorylation system. arXiv: 1904.11633 (2019).
18. Dickenstein, A., Pérez Millán, M., Shiu, A., and Tang, X.: Multistationarity in Structured Reaction Networks. Bull. Math. Biol. 81(5), 1527–1581 (2019).
19. Eithun, M., and Shiu, A.: An all-encompassing global convergence result for processive multisite phosphorylation systems. Math. Biosci. 291, 1–9 (2017).
20. Érdi, P., and Tóth, J.: Mathematical Models of Chemical Reactions: Theory and Applications of Deterministic and Stochastic Models. Manchester University Press (1989).
21. Faugère, J.-C., Moroz, G., Rouillier, F., Safey El Din, M.: Classification of the Perspective-Three-Point problem, Discriminant variety and Real solving polynomial systems of inequalities. ISSAC 2008 Proceedings, D. Jeffrey (eds), Hagenberg (2008).
22. Feinberg, M.: Foundations of Chemical Reaction Network Theory, Applied Mathematical Series, Vol. 202, Springer Nature Switzerland (2019).
23. Feliu, E., and Wiuf, C.: Enzyme-sharing as a cause of multi-stationarity in signalling systems. J. R. Soc. Interface 9 (71), 1224–1232 (2012).
24. Feliu, E., and Wiuf, C.: Simplifying biochemical models with intermediate species. J. R. Soc. Interface 10: 20130484 (2013).
25. Gatermann, K., Eiswirth, M., and Sensse, A.: Toric ideals and graph theory to analyze Hopf bifurcations in mass action systems, J. Symb. Comput. 40(6), 1361–1382 (2005).
26. Gelfand, I., Kapranov, M., and Zelevinsky,A.: Discriminants, resultants and multidimensional determinants. Birkhäuser Boston (1994).
27. Giaroli, M., Bihan, F., and Dickenstein, A .: Regions of multistationarity in cascades of Goldbeter-Koshland loops. J. Math. Biol. 78(4), 1115–1145 (2019).
28. Gnacadja, G.: Reachability, persistence, and constructive chemical reaction networks (part II): a formalism for species composition in chemical reaction network theory and application to persistence, J. Math. Chem. 49(10) , 2137–2157 (2011).

29. Gnacadja, G.: Reachability, persistence, and constructive chemical reaction networks (part III): a mathematical formalism for binary enzymatic networks and application to persistence, J. Math. Chem. 49(10), 2158–2176 (2011).

30. Gross, E., Harrington, H. A., Rosen, Z., and Sturmfels, B.: Algebraic Systems Biology: A Case Study for the Wnt Pathway Bull Math Biol. 78(1), 21–51 (2016).

31. Gunawardena, J.: A linear framework for time-scale separation in nonlinear biochemical systems, PLoS ONE 7:e36321 (2012).

32. Harrington, H. A, Mehta, D., Byrne, H., and Hauenstein, J.: Decomposing the parameter space of biological networks via a numerical discriminant approach. arXiv:1604.02623 (2016).

33. Hell, J., and Rendall, A. D.: A proof of bistability for the dual futile cycle. Nonlin. Anal. RWA 24, 175–189 (2015).

34. Hell, J., and Rendall, A. D.: Sustained oscillations in the MAP kinase cascade. Math.. Biosci. 282, 162–173 (2016).

35. Hell, J., and Rendall, A. D.: Dynamical Features of the MAP Kinase Cascade. In: Graw F., Matth´aus F., Pahle J. (eds) Modeling Cellular Systems. Contributions in Mathematical and Computational Sciences, vol 11. Springer, Cham (2017).

36. Holstein, K., Flockerzi, D., and Conradi, C.: Multistationarity in sequential distributed multisite phosphorylation networks. Bull. Math. Biol. 75(11), 2028–2058 (2013).

37. Joshi,B., and Shiu, A.: Which small reaction networks are multistationary? SIAM J. Appl. Dyn. Syst. 16(2), 802–833 (2017).

38. Mirzaev I., and Gunawardena J. : Laplacian dynamics on general graphs, Bull. Math. Biol. 75, 2118–2149 (2013).

39. Müller, S., Feliu, E., Regensburger, G., Conradi, C., Shiu, A., and Dickenstein, A.: Sign conditions for injectivity of generalized polynomial maps with appl. to chemical reaction networks and real algebraic geometry, Found. Comput. Math. 6(1), 69–97 (2016).

40. Manrai A. K., and Gunawardena J.: The Geometry of Multisite Phosphorylation, Biophysical Journal (2008), Vol. 95, 5533–5543.

41. Patel, A.L., and Shvartsman, S.Y.: Outstanding questions in developmental ERK signaling. Development 145(14) (2018).

42. Pérez Millán, and Dickenstein, A.: The structure of MESSI biological systems. SIAM J. Appl. Dyn. Syst. 17(2), 1650–1682 (2018).

43. Pérez Millán, M., Dickenstein, A., Shiu, A., and Conradi,C.: Chemical reaction systems with toric steady states. Bull. Math. Biol. 74(5) 1027–1065 (2012).

44. Qiao, L., Nachbar, R. B., Kevrekidis, I. G., and Shvartsman, S. Y.: Bistability and oscillations in the Huang-Ferrell model ofMAPK signalling. PLoS Comp. Biol. 3, 1819–1826 (2007).

45. Rubinstein, B., Mattingly, H., Berezhkovskii, A., and Shvartsman,S.: Long-term dynamics of multisite phosphorylation. Mol. Biol. Cell 27(14), 2331–2340 (2016).

46. Sadeghimanesh, AH., and Feliu,E.: The multistationarity structure of networks with interme- diates and a binomial core network. Bull. Math. Biol. 81:2428–2462 (2019).

47. Safey El Din, M.: RAGlib library, available at: https://www-polsys.lip6.fr/~safey/RAGLib/.

48. M. Sáez, C. Wiuf, E. Feliu: Graphical reduction of reaction networks by linear elimination of species. J. Math. Biol. 74(1-2), 195–237 (2017).

49. Shinar, G., and Feinberg, M.: Structural sources of robustness in biochemical reaction networks, Science 327(5971) 1389–1391 (2010), .

50. Suwanmajo, T., and Krishnan, J.: Mixed mechanisms of multi-site phosphorylation Journal of The Royal Society Interface 12(107) (2015).

51. Thomson,T., and Gunawardena, J.: The rational parameterisation theorem for multisite post- translational modification systems. J. Theoret. Biol. 261(4), 626–636 (2009).

52. Wang, L., Sontag, E.: On the number of steady states in a multiple futile cycle. J. Math. Biol. 57(1), 29–52 (2008).

Metastability: A Brief Introduction Through Three Examples

Stella Brassesco and Maria Eulalia Vares

Abstract Metastability is a very frequent phenomenon in nature. It also finds many applications in science and engineering. A noticeable basic feature is the presence of "quasi-equilibria states" and relatively sudden transitions between them. The goal of this short expository note is to discuss some aspects of the stochastic modeling of metastability, usually done through the consideration of special stochastic processes. This includes a "pathwise approach" developed since the 1980s. Thought as an invitation to the readership, three examples are quickly reviewed, starting with a class of reaction-diffusion equations subject to a small stochastic noise, for which the theory of large deviations has been a very useful tool, and further precision achieved through the help of potential theoretical techniques. We present then brief summaries of results on the Harris contact process on suitable finite graphs, and a quick discussion of stochastic dynamics for the well-known Ising model. The first can be thought as an oversimplified model for the propagation of an infection, and the second has been used in the context of magnetization. From a probabilistic analysis and technical viewpoint, the Ising model enjoys time-reversibility, which provides useful tools, while the contact process is non-reversible.

1 Introduction

Metastability is a common phenomenon in nature, with plenty of examples in physics, chemistry, and biology. Similar behavior may be also detected while studying certain phenomena in economics and social sciences. The objects of interest are systems that make transitions between quasi-equilibria states—which look like, but are not true equilibria—to stable states. Here a common example that

S. Brassesco
Instituto Venezolano de Investigaciones Científicas, Caracas, Venezuela
e-mail: sbrasses@ivic.gob.ve

M. E. Vares (✉)
Instituto de Matemática, Universidade Federal do Rio de Janeiro, Rio de Janeiro, RJ, Brazil
e-mail: eulalia@im.ufrj.br

© The Association for Women in Mathematics and the Author(s) 2019
C. Araujo et al. (eds.), *World Women in Mathematics 2018*, Association for Women in Mathematics Series 20, https://doi.org/10.1007/978-3-030-21170-7_3

we may often encounter in usual life: A very cold bottle of beer in the freezer; one takes it and the beer seems perfectly in the liquid phase, but all of a sudden it freezes while we open the bottle. Perhaps this is not such a nice event for the person who was about to drink the beer, but it gives an interesting common example of the phenomenon we want to discuss. The transition that we observe from what seemed in a normal liquid phase to the solid one is quite quick; once in the solid phase, we see it slowly getting to liquid if subject to the proper temperature.

Metastability takes place in thermodynamic systems close to a first order phase transition. We may consider starting with thermodynamic parameters that determine a state with a unique phase X and suitably changing them to those corresponding to a new phase Y. In some situations it may happen that the system does not undergo the proper phase transition and it moves instead to a sort of "quasi-equilibrium" situation with a phase X' which is very close to X. It may remain in X' for a large period of time (in the proper time scale) until a quick and unexpected change brings it to Y. Here X' corresponds to what we call "metastable state".

The classical examples include therefore supercooled liquids or vapours and supersaturated vapours or solutions: a gas at temperature below its critical value can exhibit both the liquid and the vapour phase. We start the experiment in the vapour phase and compress it, keeping the temperature fixed; at large enough pressure the thermodynamic equilibrium phase is the liquid phase, but under particular experimental situations, the system remains in the vapour phase (metastable state). This behaves essentially as a real thermodynamic equilibrium, but, either via an external perturbation or a spontaneous fluctuation, a nucleus of the "stable" phase appears and initiates a fast transition to the equilibrium liquid state.

One of the first explanations for metastability was given in the frame of the classical van der Walls-Maxwell theory, which is not compatible with the usual assumptions on the interactions between the components of the large system. There have been various efforts to describe metastability within statistical mechanics, and this might have motivated the attempt to reconcile van der Waals theory with statistical mechanics, through the introduction of Kac potentials. Nevertheless, it is by now very clear that metastability should be examined as a dynamical phenomenon. The first rigorous proposal in this direction came with the work of Lebowitz and Penrose (see [49]). They considered a deterministic dynamics given by the Liouville equation for some Hamiltonian H, with an invariant probability measure μ (the equilibrium) and proposed to describe metastable states through the conditioned measure $\mu_R = \mu(\cdot|R)$ (given by $\mu(A|R) = \mu(A \cap R)/\mu(R)$), for R a suitable subspace of the configuration space. Their choice for R was driven by three characteristics:

(i) Only one thermodynamic phase is present.
(ii) The lifetime is large, i.e. it takes a long time to exit from R.
(iii) Once it escaped from R, the return time is much longer (very unlikely to return).

Condition (ii) was expressed through a very small value of $\lambda = \frac{dp_t}{dt}$ computed at $t = 0$ and where p_t is the probability of having escaped from R by time t, if starting from μ_R, while condition (iii) can be expressed through $\mu(R)$ being very small. Lebowitz and Penrose applied this to discuss vapour-liquid transition in the case of Kac potentials.

We will use stochastic models to give a more precise description of this phenomenon, and try to answer some basic questions regarding the transition time, a characterization of what could be called "metastable state", and a description of the typical escape from the metastable state. We develop these examples mostly in the context of the so-called pathwise approach which was introduced in [18], and is based on the behavior of the empirical averages along typical trajectories. It is natural to think of Markov processes that, in spite of having a unique invariant measure, behave in the above described fashion.

We shall focus on three examples:

(1) A class of reaction-diffusion equations subject to a stochastic noise.
(2) Harris contact process on suitable finite graphs.
(3) Kinetic Ising models.

This expository paper was motivated by a lecture given by one of the authors during the 2018 World Meeting for Women in Mathematics. Nevertheless, the focus here is in metastability, a subject that appeared only tangentially in the lecture, focused on an extension of the contact process. This is by no means a detailed survey, but basically an invitation for people outside the field.

2 Stochastic Perturbations of Differential Equations

Before introducing the class of stochastic partial differential equations of our first example, let us recall some previous models in a finite dimensional setting, as they could help to understand the general picture in a simpler and more intuitive situation.

One of the earliest examples of such models was proposed by H. A. Kramers in [35] when studying the transition rates of chemical reactions. He considered, in one dimension, the equations of motion of a particle under the effect of a potential U with two wells, with local minima at the points a and b, separated by a barrier with its maximum at c, and of a Brownian motion with diffusion coefficient depending on the viscosity and the temperature. In a regime of large viscosity, the position of the particle is found to evolve approximately according to the equation

$$dX(t) = -U'(X(t))\, dt + \epsilon\, dB(t), \tag{1}$$

with ϵ a small positive parameter.

Denote by τ_ϵ the time needed by $X(t)$ to overcome the potential barrier, that is, to arrive at state b starting from a. The reasoning in [35] permits to conclude that, as the parameter $\epsilon \to 0$, and if U is smooth and $U'' \neq 0$ at the critical points, the expectation of τ_ϵ satisfies, for some constant $K > 0$,

$$E\,\tau_\epsilon \sim K\,\exp\left(\frac{2\left(U(c) - U(a)\right)}{\epsilon^2}\right). \tag{2}$$

Kramers also discusses the precise form of the prefactor K, which results to be a constant given in terms of $U''(a)$ and $U''(c)$. In this one dimensional setting, the proof that the quotient between both sides of (2) converges to 1 as $\epsilon \to 0$ follows basically from Laplace's method after considering the integrals obtained when computing the expected hitting times as solutions to the corresponding differential equations. The long time (in terms of the parameter ϵ) needed to pass from a to b, expressed in (2), is one of the characteristics of metastability.

A similar equation is obtained as a continuous limit for a class of Markov chains that arise when considering certain random dynamics associated to the Curie–Weiss model. The latter is a system of N spins $\sigma_i \in \{-1, 1\}$, whose Hamiltonian at $\sigma = (\sigma_1, \sigma_2, \cdots, \sigma_N)$ is

$$H_N(\sigma) = -\frac{1}{N}\sum_{i,j=1}^{N}\sigma_i\sigma_j - h\sum_{i=1}^{N}\sigma_i,$$

for h a constant external field that is supposed positive. As H_N is a function only of the mean magnetization $m_N(\sigma) := \frac{1}{N}\sum_{i=1}^{N}\sigma_i$,

$$H_N(\sigma) = -N\left(\left(m_N(\sigma)\right)^2 + h\,m_N(\sigma)\right),$$

the model is essentially one dimensional, the magnetization being the basic quantity. Under the corresponding Gibbs measure μ_N with inverse temperature β, defined by $\mu_N(\sigma) = \exp\left(-\beta H_N(\sigma)\right)/Z_N$, where Z_N is the normalization term, m_N is distributed in the set $\mathcal{Y}_N = \{-1 + \frac{2k}{N} : k = 0, 1, \cdots, N\}$ according to

$$\nu_N(m) = \mu_N\{\sigma : m_N(\sigma) = m\} = \frac{\binom{N}{N(m+1)/2}\,e^{\beta N(m^2 + h\,m)}}{Z_N(m)}, \qquad m \in \mathcal{Y}_N.$$

Stochastic dynamics can be associated to this system by considering Markov chains in $\mathcal{Y}_N$ such that ν_N is its unique invariant measure. This can be done in several ways, as discussed in [47, Sect. 4.3]. A particular example is presented in detail in [18], where a precise birth and death discrete time chain ξ_n is considered. For $\beta > 1$ and h sufficiently small, it behaves as $N \to \infty$ as a random walk with reflecting barriers at ± 1 and a "drift" that is a discrete approximation of the derivative of a double well function f, having a local minimum at $x^- \in (-1, 0)$, an absolute minimum at $x^+ \in (0, +1)$ and a local maximum at $x^0 \in (x^-, x^+)$. The behavior of this model, where a microscopic dynamics was prescribed, is thus very similar to that of the process $X(t)$ satisfying (1), and metastability is simple to analyse (see also [2] for the relation between metastability and cut-off).

Indeed, for each N, define m^- and m^*, both states in $\mathcal{Y}_N$ such that $x^- \in [m^-, m^- + \frac{1}{2N})$, and $x^0 + \frac{1}{N^{1/4}} \in [m^*, m^* + \frac{1}{2N})$ and consider T_N the time needed for the chain ξ_n starting at m^- to arrive at m^*. Observe that T_N is the time needed to overcome the maximum when starting close to the location of the bottom of one of the wells of f, or "tunnelling" time. The time T_N is shown in [18] to satisfy that, for each $\eta > 0$:

(i) $\quad P\left(e^{N(\Delta f - \eta)} < T_N < e^{N(\Delta f + \eta)}\right) \xrightarrow[N \to \infty]{} 1, \quad$ and $\qquad$ (3)

(ii) $\quad \dfrac{T_N}{E\,T_N} \xrightarrow[N \to \infty]{\mathcal{L}} \mathrm{Exp}(1),$ $\qquad\qquad\qquad\qquad\qquad\qquad$ (4)

where $\Delta f = f(x_0) - f(x^-)$, $\mathrm{Exp}(1)$ denotes a unit mean exponential random variable and the convergence in the last item refers to convergence in law. A third statistical property of the chain ξ_n is the stability of the temporal means of trajectories over long times R_N (again, with respect to the small parameter $1/N$ in this model), which, when starting from m^- and before T_N behave as if there were only one well.

To state it precisely, let us consider the empirical process of time averages over intervals of length $R > 0$:

$$A_R(s) = \frac{1}{R} \sum_{n=[s]+1}^{[s]+R} \delta_{\xi_n},$$

where δ_x is the measure concentrated at x. For each Borel set $B \subset [-1, 1]$ denote $A_R(s; B) = A_R(s)(B)$, that is the proportion of time that the chain spends in B during the time interval $\big[[s]+1, [s]+R\big]$. Then, for suitable sequences $R_N \to \infty$ such that $R_N/e^{N\Delta f} \to 0$, for instance $R_N = e^{\alpha N}$ for $\alpha \in (0, \Delta f)$, and if $D \subset [-1, 1]$ is an open set that contains m^-, for each $\rho > 0$:

(iii) $\quad P_{m^-}\left(\inf_{s < T_N - R_N} A_{R_N}(s; D) > 1 - \rho,\ T_N > R_N\right) \xrightarrow[N \to \infty]{} 1.$ $\qquad$ (5)

The subindex in P_x indicates that the initial condition is x.

A more complete analysis of the Curie–Weiss model can be obtained by considering, instead of the magnetization chain, the full evolution of the spin system through a spin flip dynamics for μ_N (see [36]).

In [18], the authors proposed the characterization of the phenomenon of metastability for systems with stochastic dynamics in terms of the three features (i), (ii) and (iii) presented above. That approach, or "pathwise approach" was applied in [29] to a class of processes given as solutions to stochastic differential equations of the type (1) in dimension $d > 1$, where it is shown that they also exhibit the phenomenon of metastability as characterized by the properties corresponding to (i), (ii) and (iii) above in that case. More precisely, consider $X(t)$ the diffusion process in $\mathbb{R}^d$ given as the solution in the Itô sense of the equation

$$dX(t) = -\nabla U\big(X(t)\big)\,dt + \epsilon\,dB(t). \tag{6}$$

The potential $U : \mathbb{R}^d \to \mathbb{R}$ is a C^2 double well function having exactly three critical points a, b and c with $U(b) < U(a) < U(c)$ such that a and b are local minima and c is a saddle point. Suppose furthermore that the determinant of the Hessian matrix at the critical points does not vanish, that the standard conditions for the equation (6) to have a unique strong solution hold, and that U increases sufficiently fast as $|x| \to \infty$. Under those assumptions, as $\epsilon \to 0$ (i), (ii) and (iii) hold for X_t. The corresponding tunnelling time T_ϵ is here the time needed to arrive at a neighborhood of b starting from a. In particular, its expectation satisfies

$$\epsilon^2 \log E\, T_\epsilon \xrightarrow[\epsilon \to 0]{} 2\big(U(c) - U(a)\big),$$

and the corresponding R_ϵ in (iii) can be taken as $R_\epsilon = \exp(\alpha/\epsilon^2)$, for any $0 < \alpha < 2\big(U(c) - U(a)\big)$. The main ingredient in the proofs in [29] are the large deviations estimates of Freidlin and Wentzell, [28], which permit to obtain asymptotic estimates for the exit times also in cases where the drift coefficient in (6) has a much more complicated structure of critical points. Freidlin and Wentzell theory yields asymptotic logarithmic equivalence for the tunnelling times, as stated above. More recently, precise estimates of the type provided by Kramers including the corresponding prefactor K as in (2), which is known as Eyring–Kramers formula, were obtained for the expectation of the tunnelling times in this type of models with the aid of potential theoretical tools. We refer to [9] or [10] for further details and precise results on this approach.

Let us turn now to consider stochastic perturbations of the partial differential equation

$$\partial_t u = \Delta u - V'(u), \tag{7}$$

where Δ denotes the Laplacian and V' is the derivative of a double well polynomial. The above is a well studied equation that appears in several contexts that have in common the coexistence of two phases. It is for instance a phenomenological model for the formation and evolution of interfaces in binary alloys. In that case, u represents the relative concentration of one of the components. It is referred to as well as Allen–Cahn equation in the literature, after [1], where this phenomenon is studied. This type of reaction-diffusion equation appears also as the macroscopic limit for a system of interacting spins obtained by superposing a spin flip dynamics of the type considered in Sect. 4 (reaction term) and a fast stirring process (the diffusion term), responsible for some sort of propagation of chaos property, with the validity of a deterministic limit for the magnetization, the fluctuations being described by a suitable space-time Gaussian process. This family, usually called "Glauber+Kawasaki process", was introduced in [19]. In spite of partial results, the metastability description, as pointed out in the notes at the end of Ch. 5 in [47] appears not yet completely answered. (See e.g. [12, 26, 33], and [10].)

Let us recall the main features of the equation (7) in the case $x \in [0, L]$. It is presented in [32] as an example of a dynamical system in an infinite dimensional space, whose structure is analogous to that considered before in $\mathbb{R}^d$.

To simplify the exposition, we consider Dirichlet boundary conditions $u(0, t) = u(L, t) = 0$ and $V(u) = \frac{u^4}{4} - \frac{u^2}{2}$. The equation is known to have a bounded continuous solution for each continuous initial condition φ. The notation $u(x, t; \varphi)$ is used when the initial condition is to be emphasized. One may think $u(\cdot, t; \varphi)$ as a function of t taking values in some convenient subspace of $C[0, L]$, and write (7) in a form that resembles the form already considered in the finite dimensional situation:

$$\partial_t u = -\frac{\delta S(u)}{\delta u}, \tag{8}$$

where the "potential" now is the functional defined for $\psi \in H^1[0, L]$ by

$$S(\psi) = \int_0^L \{\frac{1}{2}(\psi')^2 + V(\psi)\} \, dx,$$

and the derivative is a functional one. We refer to [14] and [32] for a detailed analysis of the functional S and the resulting dynamical system given by (8). Its main features can be summarized as follows:

- If $N\pi < L \leq (N + 1)\pi$, then

    The functional S has $2N + 1$ critical points: $\pm m_1, \pm m_2, \cdots, \pm m_N$ and 0. They are the stationary solutions of (7).
    Each $\pm m_n$ has n nodes, and $S(\pm m_1) < S(\pm m_2) < \cdots < S(\pm m_N) < 0 = S(0)$. In particular, if $L > \pi$, m_1 and $-m_1$ are the two minimizers of S, and the rest are saddle points.

- $S(u(\cdot, t))$ is a decreasing function of t, and $u(\cdot, t) \to m$ as $t \to \infty$, for some critical point m.
- The minima $\pm m_1$ are asymptotically stable in the sense of Lyapunov and their domains of attraction $\mathcal{D}^{\pm}$ are open sets, when considering the sup norm in $C[0, L]$.

We suppose that $L > \pi$ in what follows; the general picture is thus similar to that of the finite dimensional case: the dynamics is that of a gradient system under the effect of a potential with two minima. It has several saddle points, which are located in the separatrix of the domains of attraction of the minimizers.

It results then natural to consider the equations (7) under the effect of small noise, to account for the internal fluctuations and for possible small neglected terms in the derivation of the equations when obtained from phenomenological models. The perturbation given by an additive space time white noise $\dot{W}$ is an interesting first instance, that was considered by W. Faris and G. Jona–Lasinio in the pioneering paper [27]:

$$\begin{cases} \partial_t u(x, t) = \partial_{xx} u(x, t) - V'(u(x, t)) + \epsilon \dot{W}, \\ u(0, t) = u(L, t) = 0. \end{cases} \tag{9}$$

As remarked in [27], (9) can be seen as a dynamical model associated with the energy functional S whose invariant measure is, at least formally, proportional to $\exp\left(-\frac{2}{\epsilon^2} S(u)\right)$, what motivates the study of infinite dimensional diffusions.

To give a rigorous meaning to equation (9), an equivalent integral equation is derived in terms of the integral operator $(\partial_t - \partial_{xx})^{-1}$. The term resulting from applying this operator to $\dot{W}$ is a Gaussian process with continuous Hölder covariance in both variables (x, t), having thus continuous paths with probability one. A standard fixed point argument yields uniqueness and existence of the solution as a process $u^{(\epsilon)}(x, t; \varphi)$ for each $\varphi \in C[0, L]$, with paths in $C([0, L] \times [0, T])$ a.s. for a small T, result that can be extended to any $T > 0$ using the particular features of the equation. It is seen to be a Markovian process over $C[0, L]$. The detailed proofs are presented in [27], where large deviation estimates are obtained for $u^{(\epsilon)}$, as $\epsilon \to 0$. Those generalize the Wentzell and Friedlin estimates for small perturbations of dynamical systems in $\mathbb{R}^d$. The computation of the probabilities of small deviations from the deterministic trajectories over finite time intervals are used in [27] to study the tunnelling, that is, the event that a trajectory starting close to one of the minimizers passes to be close to the other minimizer. In particular, estimations for the time needed for that event are obtained, which suggest that the passage takes place through a small neighborhood of the saddles with lower potential, that is, $\pm m_2$. In subsequent works the picture was indeed completed, to show that $u^{(\epsilon)}$ has a metastable behavior similar to the finite dimensional models considered in [29]. We need to introduce some definitions before stating the corresponding results and references.

We suppose without further mention that the considered functions in $C[0, L]$ satisfy Dirichlet boundary conditions, and that the norm in this space is the sup norm. For each $\psi \in C[0, L]$ let $B_c(\psi)$ denote the ball in that space centred at ψ with radius c. Let T_ϵ be the time of arrival at a neighborhood of $-m_1$ by the process $u^{(\epsilon)}$:

$$T_\epsilon = \inf\{t \geq 0 : u^{(\epsilon)}(\cdot, t; \varphi) \in B_c(-m_1)\}.$$

It depends on the initial condition, which is (sometimes) included in the notation as sub index in the corresponding probability.

Define the normalization term γ_ϵ implicitly by

$$P_{m_1}\{T_\epsilon > \gamma_\epsilon\} = \frac{1}{e},$$

and let $\Delta = S(m_2) - S(m_1)$. Then, metastability holds for the process $u^{(\epsilon)}$, as detailed below:

(i) Given any $\eta > 0$,

$$P_\varphi\left(e^{\left(\frac{2\Delta-\eta}{\epsilon^2}\right)} < T_\epsilon < e^{\left(\frac{2\Delta+\eta}{\epsilon^2}\right)}\right) \to 1 \text{ as } \epsilon \to 0,$$

uniformly for $\varphi \in B_c(m_1)$, for sufficiently small c.

(ii) $E_{m_1}\frac{T_\epsilon}{\gamma_\epsilon} \to 1$ as $\epsilon \to 0$ and $P_{m_1}\left(\frac{T_\epsilon}{\gamma_\epsilon} > t\right) \to e^{-t}$ as $\epsilon \to 0$.

(iii) There exists a sequence $R_\epsilon \to \infty$ and $\frac{R_\epsilon}{\gamma_\epsilon} \to 0$ as $\epsilon \to 0$ such that for each $\delta > 0$ and $f : C[0, L] \to \mathbb{R}$ a continuous and bounded function,

$$P_\varphi\left(\sup_{0 \leq t \leq T_\epsilon - 2R_\epsilon} \left|\frac{1}{R_\epsilon}\int_t^{t+R_\epsilon} f\left(u^{(\epsilon)}(\cdot, s)\right) - f(m_1)\,ds\right| > \delta\right) \to 0 \text{ as } \epsilon \to 0.$$

Item (i) follows from the large deviation estimates obtained in [27]. The convergence of the normalized tunnelling time stated in (ii) is proved in [39] using a coupling of trajectories starting at different points and the particular features of S. The statement (iii) concerning the stability of the temporal averages is proved in [11], where it is observed that the sequence R_ϵ can be indeed taken as $R_\epsilon = \exp(\alpha/\epsilon^2)$, for $0 < \alpha < 2\left(S(m_2) - S(m_1)\right)$. As suggested by the previous results, it is proved also that the tunnelling occurs trough a neighborhood of $\pm m_2$, which are the saddles with lowest value of S.

Accurate estimates of the tunnelling times, that result similar to the Eyring–Kramers formula, were deduced in [3] as a result of approximating the equation by a system of coupled diffusions for which the potential approach can be used, and yield the prefactor term.

More recently, motivated by reported behavior of climate models, small perturbations of dynamical systems with non Gaussian Lévy noise have been considered, both in a finite and infinite dimensions. The noise does not have exponential moments in this case, and thus the large deviation estimates are not valid, and different techniques are needed. The normalized exit times from neighborhood of the attractors are seen to converge to an exponential random variable as well in this case, but the magnitude of the expectation results to be of order $\epsilon^{-\rho}$, with $\rho > 0$ depending on the precise characteristics of the Lévy noise, in contrast with the exponentially long times of the Gaussian case. For more results concerning this models, we refer to [20, 21] and [34], and the references therein.

3 Harris Contact Process

Introduced by Harris in [31], the contact process is a very simple stochastic model for the propagation of an infection. Each vertex of a locally finite connected graph

$G = (V, \mathcal{E})^1$ is identified with an individual, which can be infected or healthy. An infected individual may infect, with rate λ, each of its healthy neighbors, or may get healthy, with rate 1. This defines a Markov process on the state space of all possible configurations ξ, where we set $\xi(x) = 1$ if x is infected and $\xi(x) = 0$ if x is healthy. A configuration is naturally identified with the set of infected individuals. In the most studied case, the graph is determined by the cubic lattice $V = \mathbb{Z}^d$ with an edge $e = \{x, y\}$ between vertices x and y if and only if $\|x - y\|_1 = 1$, where $\| \cdot \|_1$ refers to the usual ℓ_1-norm in $\mathbb{R}^d$. In this case we write $\mathbb{E}^d$ for the set of edges.

This process has been extensively studied also on more general graphs, including long range interactions on $\mathbb{Z}^d$, trees, or a class of random graphs. For the general theory we refer to [37, Ch. VI] and [38, Part I], as the two classical general references on the subject of interacting particle systems. Regarding dynamics on random graphs, a monograph to be mentioned at this point is [23]. Other references will be given below.

For the rigorous construction of the process, one may appeal to the standard semigroup theory through an application of Hille-Yosida theorem, as in [37]. It is nevertheless simpler and very useful to describe its construction via a "Harris system", as we do now. On a suitable probability space $(\Omega, \mathcal{A}, P)$ we take:

(a) $\{N_e : e \in \mathcal{E}\}$, a system of i.i.d. Poisson processes of rate λ.
(b) $\{N_x : x \in V\}$, a system of i.i.d. Poisson processes of rate 1, assumed to be independent of the system in (a).

Given these processes, that constitute what we name a *Harris system*, for times $0 \le s < t$, and $x, y \in V$, a path γ from (x, s) to (y, t) is a càdlàg[2] function from $[s, t]$ to V for which there exist times $t_0 = s < t_1 < \cdots < t_k = t$ and sites $x_0 = x, x_1, \ldots, x_{k-1} = y$ in $\mathbb{Z}^d$ such that $\gamma(u) = x_i$ for $u \in [t_i, t_{i+1})$ and

- $N_{x_i} \cap [t_i, t_{i+1}] = \emptyset$ for $i = 0, \ldots, k - 1$;
- $\{x_i, x_{i+1}\} \in \mathcal{E}$ for $i = 0, \ldots, k - 2$;
- $t_i \in N_{\{x_{i-1}, x_i\}}$ for $i = 1, \ldots, k - 1$.

For $A \subset V$, the contact process starting from the initial configuration A is defined as

$$\xi_t^A = \{y : \text{ there exists a path from } (x, 0) \text{ to } (y, t) \text{ for some } x \in A\}. \tag{10}$$

(We omit the initial configuration from the notation when this is clear from the context.) The big interest in this process has to do with the fact that it presents a dynamical phase transition, i.e. its dynamical and ergodic behavior change drastically as one varies the parameter λ. If $(V, \mathcal{E}) = (\mathbb{Z}^d, \mathbb{E}^d)$, there exists a critical value of the infection parameter $\lambda_c(d) \in (0, \infty)$ so that a non-trivial invariant

---

[1] V is the set of vertices and $\mathcal{E}$ denotes the set of unordered edges.
[2] Right continuous, with left limits.

measure μ_λ exists if and only if $\lambda > \lambda_c(d)$ (of course the Dirac point mass at the empty configuration is always an invariant measure).

The contact process has various special properties, including monotonicity (in the stochastic sense) in the initial configuration (also called attractiveness) and in the rate λ. Indeed, the process has the following additivity property: $\xi_t^{A \cup B} = \xi_t^A \cup \xi_t^B$ for all A, B and t, which of course implies the attractiveness. The measure μ_λ is the limit (in law) of $\xi_t^{\mathbb{Z}}$ as $t \to \infty$. The process also enjoys the self-duality property:

$$P(\xi_t^A \cap B \neq \emptyset) = P(\xi_t^B \cap A \neq \emptyset), \tag{11}$$

for all A, $B \subset V$ and all $t > 0$. All these properties are easily verified if we consider the construction via the Harris system. One has $\lambda_c(d) = \inf\{\lambda: P(\xi_t^{\{0\}} \neq \emptyset \ \forall t) > 0\}$ and

$$\mu_\lambda\{\eta: \eta(x) = 1 \text{ for some } x \in B\} = P(\tau^B = \infty), \tag{12}$$

for each finite set B, where

$$\tau^B = \inf\{t: \xi_t^B = \emptyset\} \text{ (setting } \inf \emptyset = \infty). \tag{13}$$

In particular

$$\rho_\lambda := \mu_\lambda\{\xi: \xi(0) = 1\} = P(\tau^{\{0\}} = \infty). \tag{14}$$

If $\lambda > \lambda_c(d)$, the measures μ_λ and $\delta_\emptyset$ are the unique extremal invariant measures of the contact process on $\mathbb{Z}^d$, and for any starting configuration the process converges in law to a mixture of them. This is the so-called Complete Convergence Theorem, proved by Durrett for $d = 1$ (see [22] or Chapter VI in [37]). The proof relies on a renormalization (block) argument. It is easy to understand why this theorem should be expected in this case: if we consider the two processes $\xi_t^{\{0\}}$ and $\xi_t^{\mathbb{Z}}$, built on the same Harris system, we see that on the event $\{\tau^{\{0\}} = \infty\}$ they coincide inside a random interval $[\ell_t, r_t]$ such that $r_t/t \to \alpha > 0$ and $\ell_t/t \to -\alpha$ almost surely as $t \to \infty$, and where α is a positive constant, i.e.

$$\xi_t^{\{0\}}(x) = \xi_t^{\mathbb{Z}}(x) \quad \forall x \in [\ell_t, r_t].$$

To see why this is true, observe first that two paths that cross in our Harris system must meet, because we are taking $d = 1$ and the infections are of nearest neighbor character. This implies that on the event $\{\tau^{\{0\}} > t\}$ we have

$$\max \xi_t^{\{0\}} = \max \xi_t^{\mathbb{Z}_-} =: r_t \text{ and } \min \xi_t^{\{0\}} = \min \xi_t^{\mathbb{Z}_+} =: \ell_t,$$

where $\mathbb{Z}_- = \{x \in \mathbb{Z}: z \leq 0\}$ and $\mathbb{Z}_+ = \{x \in \mathbb{Z}: z \geq 0\}$. The subadditive ergodic theorem provides the a.s. limit mentioned before. As we guess from the above considerations, for any initial non-empty configuration A, the law of ξ_t^A converges to $P(\tau^A = \infty)\mu_\lambda + P(\tau^A < \infty)\delta_\emptyset$.

The Complete Convergence Theorem holds for any $d \geq 2$. This follows from dynamical renormalization methods developed in [6] (built also on [4], which deals with unoriented percolation), which play a crucial role for the verification of metastability as described below. As before, renormalization refers to a consideration of (well-connected) space-time blocks, but now the blocks are randomly and dynamically constructed, which yields more flexibility. The dynamical renormalization developed in [6] provides a finite volume criteria (therefore continuous in λ) for the existence of infinite paths or survival. In particular, this implies that the process dies out at criticality, for any $d \geq 1$.

If we restrict the contact process to a finite spatial box $\Lambda_n = [-n, n]^d$, i.e. restrict the Harris system to corresponding subgraph with vertices in Λ_n, we end up with a finite Markov chain ξ_n and

$$T_{\Lambda_n} := \inf\{t : \xi_{n,t} = \emptyset\} \tag{15}$$

is a.s. finite, for any initial configuration, i.e. $\delta_\emptyset$ is the unique invariant measure. Nevertheless, the behavior of this finite volume contact process is very different according to $\lambda < \lambda_c(d)$ or $\lambda > \lambda_c(d)$. Assuming, to fix ideas, that all individuals are initially infected, in the subcritical case, $T_{\Lambda_n}/\log n$ converges to a constant in probability (see [16] and [24]). On the other hand, if $\lambda > \lambda_c(d)$, the time of extinction grows exponentially in n, i.e.

$$\lim_{n \to \infty} \frac{1}{n} \log E\, T_{\Lambda_n} = c, \tag{16}$$

for a positive and finite constant c (see [25] and [41]). Moreover, for large n the process undergoes metastability before its extinction, the metastable state being approximated by the restriction of μ_λ to $\{0, 1\}^{\Lambda_n}$. This was formulated as follows:

Theorem 1 *Assume $\lambda > \lambda_c(d)$ and start with all individuals in Λ_n infected at time 0. Then (writing simply T_n for T_{Λ_n}):*

(i) *The sequence $T_n/E\, T_n$ converges in distribution to a unit mean exponential random variable.*

(ii) *There exist $R_n \to \infty$ with $R_n/E\, T_n \to 0$ as $n \to \infty$ so that for each $\epsilon > 0$ and each $f : \{0, 1\}^{\Lambda_n}$ which depends on finitely many coordinates,*

$$\lim_{n \to \infty} P\left(\sup_{s < T_n - 2R_n} |A_{R_n}^n(s, f) - \mu_\lambda(f)| \leq \epsilon, T_n > R_n \right) = 1, \tag{17}$$

where

$$A_R^n(s, f) = \frac{1}{R} \int_s^{s+R} f(\xi_{n,u})\,du. \tag{18}$$

Part (i) simply says that the extinction time is asymptotically unpredictable, while part (ii) says that the system "thermalizes" at a state close to μ_λ for large values of n. Theorem 1 in the case $d = 1$ goes back to [18] and [52]. The proof of (i) and the validity of (16) in the higher dimensional case are due to Mountford [40], who developed, exploiting the Bezuidenhout–Grimmett renormalization, a totally new way of looking at the regeneration property that is behind the result. This is contained in the proposition below, which provided the main ingredient for the loss of memory included in part (i) of Theorem 1, stating that with overwhelming probability, and in a time scale much shorter than that in which the finite process gets extinct, the process forgets its initial configuration. With suitable quantitative estimates this was used in [56] to get (ii) for $d \geq 2$.

Notation In the following we use $\underline{1}$ to denote the (initial) configuration with everyone infected, $\xi(x) = 1$ for all x.

Proposition 1 (Regeneration [40]) *There exist sequences $\{a_n\}$ and $\{b_n\}$ both tending to infinity, so that*

(a) $b_n/a_n \to \infty$;

(b) $\sup_{\eta \in \{0,1\}^{\Lambda_n}} P(\xi_{a_n}^{\underline{1}} \neq \xi_{a_n}^\eta, T_n^\eta > a_n) \to 0$;

(c) $P(T_n^{\underline{1}} > b_n) \to 1$.

It is known since [48] and [57] that the contact process on more general graphs has a richer phase transition structure. One good example is the infinite d-ary tree $\mathbb{T}^d$ (all vertices have degree $d + 1$). The process can survive globally, i.e. $P(\tau^{\{x\}} = \infty) > 0$ and yet the infection does not return to site x infinitely often, $P(\limsup_{t \to \infty} \xi_t^{\{x\}} = 1) = 0$. This implies the existence of two different critical parameters $0 < \lambda_1 = \lambda_1(\mathbb{T}^d) < \lambda_2 = \lambda_2(\mathbb{T}^d) < \infty$ (which coincided in the case of $\mathbb{Z}^d$):

$$\lambda_1 = \sup\{\lambda : P(\tau^{\{x\}} < \infty) = 1\},$$

$$\lambda_2 = \inf\{\lambda : P(\limsup_{t \to \infty} \xi_t^x = 1) > 0\}.$$

It is also known that at λ_1 the process dies out, and that at λ_2 it survives only globally. The type of Complete Convergence Theorem mentioned earlier holds whenever $\lambda > \lambda_2$ for $\mathbb{T}^d$; in particular, for $\lambda > \lambda_2$ the process has only two extremal invariant measures, as in the whole supercritical region for $\mathbb{Z}^d$, a result initially proven in [58]. See also [51] for a different proof and [50] for a description of an even much richer ergodic behavior for the contact process on a more general class of graphs.

It is natural to ask about the extension of the previous results on metastability to the context of $\mathbb{T}^d$. Stacey [57] considered the restriction of the contact process to $\mathbb{T}_n^d$, the rooted d-ary tree of height n: the root has degree d, the leafs have degree 1, other vertices have degree $d + 1$ (like a branching tree with d descendants stopped at the nth generation). His results indicated a behavior similar to that of the contact process on Λ_n. In particular, if $\lambda < \lambda_2$, and if we start with everyone infected, the extinction time $T_{\mathbb{T}_n^d}$ is of order n, which is the exact analogue of the corresponding result in Λ_n i.e. $T_{\mathbb{T}_n^d} / \log |\mathbb{T}_n^d|$ tends in probability to a suitable constant, as $n \to \infty$. The results in [57] for the case $\lambda > \lambda_2$ were complemented in [17] and then in [42]. One of the related results in these papers is the following:

Theorem 2 ([17, 42]) *Consider the contact process on* $\mathbb{T}_n^d$ *with* $\lambda > \lambda_2$ *as before, and start with all individuals in* $\mathbb{T}_n^d$ *infected at time* 0. *Then:*

(a) *There exists a finite, positive constant c so that* $\lim_{n \to \infty} \frac{1}{|\mathbb{T}_n^d|} \log E \, T_{\mathbb{T}_n^d} = c$.

(b) *As* $n \to \infty$, $T_{\mathbb{T}_n^d} / E \, T_{\mathbb{T}_n^d}$ *converges in distribution to a unit mean exponential random variable.*

Remark Investigating a larger class of finite random graphs with bounded degree, Theorem 1.2 in [44] sheds new insight into the breaking point between the logarithmic and exponential behaviors, as in parts (a) and (b) in Theorem 2. They consider a sequence of random graphs for which the breaking point is $\lambda_1(\mathbb{T}^d)$, characterized as the limiting graph (in some sense) of the finite graphs under consideration. This fits well to the result in Theorem 2, in which case the limiting graph is the so called canopy tree, for which $\lambda_1 = \lambda_2 = \lambda_2(\mathbb{T})$

There is very active research on the behavior of the contact process on a class of (finite) random graphs (V_n, E_n). In the spirit of part (ii) of Theorem 1, in [43], the authors study the empirical density of the process at times t_n which are much shorter than the typical extinction time T_{V_n}. Another interesting situation comes from the consideration of the contact process on the preferential attachment graph (see [5, 13]) and other finite graphs with power law degree distribution. In these cases, the metastability manifests itself through the behavior of the extinction time no matter how small the infection parameter $\lambda > 0$ is (i.e. $\lambda_c = 0$); see [15].

4 Kinetic Ising Model

A well known example of metastability that arises near a phase transition has to do with ferromagnetic systems below the Curie critical temperature T_c and the coexistence of two different phases for a null magnetic field. In a real ferromagnet one frequently observes the well known hysteresis loop: if we start with a small field $h < 0$ and slowly increase it to zero and to small positive values, the magnetization naturally increases but remains negative, in a situation of apparent equilibrium. This remains valid up to a certain value h^*, called coercive field, for which there is a decay to a stable state with positive magnetization. Reversing the procedure one observes the analogous behavior with reversed signs. Natural questions are: can one describe this "metastable" magnetization? Can we say something about the time it takes for the transition to stability?

To discuss this in a mathematical precise form, one naturally considers stochastic models for a ferromagnet. The simplest such example is the famous Ising model, here considered on the two-dimensional lattice $(\mathbb{Z}^2, \mathbb{E}^2)$, with $\mathbb{E}^2$ as in the previous section. In the Ising model, to each vertex x in the graph we associate a random variable σ_x that can take only two values $+1$ or -1, simulating a spin with two possible orientations.

Given a finite set $\Lambda \subset \mathbb{Z}^2$ we consider an energy function $H_{\Lambda,h}(\sigma)$ on the space of all possible configurations $\Sigma_\Lambda = \{-1, +1\}^\Lambda$. In order to model a ferromagnet, the energy favors the alignment of spins at nearest neighbor vertices. We also assume the existence of a (constant) magnetic field which also favors alignment in a given direction. We consider two situations: (i) a fixed configuration $\bar{\sigma}$ outside Λ; (ii) periodic boundary conditions, i.e. $\Lambda = [-N, N]^2$ taken as a two dimensional discrete torus. In the first case we write

$$H_{\Lambda,h,\bar{\sigma}}(\sigma) = -\frac{1}{2} \sum_{\substack{x,y \subset \Lambda, \\ \|x-y\|_1 = 1}} \sigma(x)\sigma(y) - \frac{h}{2} \sum_{x \in \Lambda} \sigma(x) - \frac{1}{2} \sum_{\substack{x \in \Lambda, y \notin \Lambda, \\ \|x-y\|_1 = 1}} \sigma(x)\bar{\sigma}(y), \quad (19)$$

where $\bar{\sigma}$ represents a boundary condition.[3] In the periodic case, or when describing something that applies to both cases, we simply write $H_{\Lambda,h}$. The energy is modulated by a parameter $\beta > 0$, so that $1/\beta$ represents the temperature, thus defining the corresponding finite volume Gibbs measure:

$$\mu_{\Lambda,\beta,h,\bar{\sigma}}(\sigma) = \frac{e^{-\beta H_{\Lambda,h,\bar{\sigma}}(\sigma)}}{Z_{\Lambda,\beta,h,\bar{\sigma}}}, \quad \sigma \in \Sigma_\Lambda,$$

with $Z_{\Lambda,\beta,h,\bar{\sigma}}$ the normalizing constant that turns $\mu_{\Lambda,\beta,h,\bar{\sigma}}$ into a probability measure on Σ_Λ. The interest of the Ising model comes from its simplicity and the fact that it presents a phase transition when $h = 0$: there exists $\beta_c \in (0, \infty)$ so that for $\beta > \beta_c$ (low temperature) there are multiple infinite volume Gibbs measures corresponding to the interaction energy described before; the pure phases are represented by the extremal Gibbs measures.

This phase transition can also be formulated by ignoring external conditions, but looking at what happens when we vary h. When $\beta > \beta_c$ there is a spontaneous magnetization: the average value of the spin at the origin in the limit as $N \to \infty$ is a function $m(\beta, h)$ that tends to $m_+(\beta) > 0$ ($-m_+(\beta)$) as h tends to zero through positive (negative, respectively). This was originally obtained by Peierls (1936) and the exact value of β_c for the Ising model in $\mathbb{Z}^2$ was obtained by Onsager (1944), proving that $m_+(\beta_c) = 0$ in this case.

Metastability is a genuinely dynamical phenomenon. In order to describe it within the same frame of the other examples, we fix a stochastic process that has the above Gibbs measure $\mu_{\Lambda,\beta,h}$ as its unique invariant measure. Usually called

---

[3] We write $h/2$ instead of h in (19) just to match with the notation in [30, 55].

kinetic Ising models, there are plenty of choices, the simplest being the spin flip Markov chains, i.e. the transitions are always of the type σ to σ^x, where

$$\sigma^x(y) = \begin{cases} -\sigma(x) & \text{if } y = x, \\ \sigma(y) & \text{if } y \neq x, \end{cases}$$

leading to an irreducible continuous time Markov chain on Σ_Λ, with infinitesimal generator

$$Lf(\sigma) = \sum_{x \in \Lambda} c(\sigma, \sigma^x)\{f(\sigma^x) - f(\sigma)\},$$

where $c(\sigma, \sigma^x)$ are the flip rates.

One example is the so-called Metropolis chain, with jump rates

$$c(\sigma, \sigma^x) = e^{-\beta\left[H_{\Lambda,h}(\sigma^x) - H_{\Lambda,h}(\sigma)\right]_+}, \qquad \sigma \in \Sigma_\Lambda, \qquad x \in \Lambda,$$

where $[a]_+ = \max\{a, 0\}$. Since these rates satisfy

$$\mu_{\Lambda,\beta,h}(\sigma)c(\sigma, \sigma^x) = \mu_{\Lambda,\beta,h}(\sigma^x)c(\sigma^x, \sigma),$$

$\mu_{\Lambda,\beta,h}$ is the unique invariant probability measure for the chain (indeed the dynamics under $\mu_{\Lambda,\beta,h}$ is time reversible).

The pioneer work dealing with the pathwise description for metastability in the context of kinetic Ising models came from Neves and Schonmann (see [45]). They considered the situation of fixed N, with periodic boundary condition on Λ_N, $h > 0$, and considered the low temperature limit $\beta \to \infty$.[4] The interesting range is $0 < h < 2$. In this range, in [45, 46, 53] the authors give a precise description of the *critical droplet*, the typical time and typical pattern of metastability in this case, starting e.g. from the configuration with all spins down -1. Among other things, they prove the analogue of Theorem 1. The intuition is simple: since $\beta \to \infty$, and we have a positive magnetic field, the finite volume Gibbs measure tends to δ_{+1} the Dirac point mass at the configuration $\sigma(x) = 1$ for all x which we denote by $+1$. The "critical droplet" has a well defined shape: a rectangle of pluses with sides $L \times (L - 1)$ with a $+1$ attached to one of the largest sides, where $L = \lceil 2/h \rceil$ (and of course we need N suitably large), and for technical reasons assume that $2/h$ is not an integer. The picture in this case is very similar to that of Wentzell-Freidlin regime with a double well potential: before overcoming the barrier of a critical droplet the process quickly returns to a neighborhood of -1 where it spends most time. Once a square $L \times L$ is formed there is a quick approach to $+1$.

---

[4]This is not the situation described in the previous page where $h \to 0$ and the volume must grow. It is much simpler and opened the door to a huge amount of work in the mathematical description.

Since [45], there has been a huge amount of work by several groups of researchers, and we refer to the monographs [47] (Ch. 6 and 7, up to 2004) and [10] (more recent), as well as to the introduction of [7] for reviews and more detailed information. These works employ different techniques, including large deviations in the spirit of Freidlin-Wentzell theory or tools from potential theory. In some cases a great deal of sophisticated analysis is demanded. In some of these situations, like in the example we just discussed, the metastable state is quite concentrated in a small subset of the configuration space, but this is not the case in some other regimes, as the one described below.

A fundamental contribution came with Schonmann and Shlosman. In [55] the authors considered the infinite volume kinetic Ising model in the vicinity of the phase-coexistence line, i.e. $\beta > \beta_c$ and very small magnetic field $h > 0$. Let μ_- and μ_+ denote the two extremal Gibbs measures for the given β and $h = 0$, with $\mu_+(\sigma(x)) = m_+(\beta) > 0$. They investigated the approach to equilibrium for the dynamics, with an initial measure that is stochastically smaller than μ_- in the FKG sense. A number $\Delta_c \in (0, +\infty)$ is determined so that for $h > 0$ small and $t = e^{\Delta/h}$ with $\Delta < \Delta_c$ the average values of local observables at time t are close to those of a metastable state, near the minus phase μ_- (in spite of the presence of a positive magnetic field). On the other hand, if $\Delta > \Delta_c$ (and $h > 0$ small) the averages at time $t = e^{\Delta/h}$ are close to those in the plus phase μ_+. The value of Δ_c is determined through a remarkable expression in terms of the equilibrium quantities:

$$\Delta_c = \Delta_c(\beta) = \frac{\beta(w_\beta)^2}{12 m_+(\beta)}, \tag{20}$$

where w_β is the surface tension of a Wulff shape of volume one (see [55]) and $m_+(\beta)$ is the spontaneous magnetization mentioned before.

In order to complete the metastability picture within the pathwise approach it would be interesting to have a result similar to that in Theorem 1, describing an asymptotically exponential tunnelling time. For this it is necessary to restrict the process to a finite volume that suitably grows as h decreases. This matter has been considered in [30] for the dynamics with fixed minus external condition and $\Lambda_h \subset \mathbb{Z}^2$ with volume (cardinality) of order $(C/h)^2$. The result is proven when Λ_h is obtained as the intersection with $\mathbb{Z}^2$ of a Wulff shape of area $(C/h)^2$ in $\mathbb{R}^2$ for C large enough. In this simpler context it is possible to define a suitable metastable state, whose typical configurations have small plus-phase droplets in a see of minus spins, and it is possible to verify that the transition to equilibrium happens after an approximate exponential time. For a precise statement we refer to [30]. The metastable state is defined in terms of a suitably restricted Gibbs measure, i.e. $\mu_{\Lambda_h,h,-}$ conditioned on a convenient set $\mathcal{R}$ that does not allow for large plus-phase contours. A crucial point consists in showing that the dynamics restricted to $\mathcal{R}$ relaxes to a vicinity of the conditioned Gibbs measure in a time scale that is much shorter than the one needed to exit. Here the crucial estimates come from [8] (see also [7]). We should remark that the "escape" or "tunnelling" time $T_{\beta,h}$ that plays the important role is not exactly the exit time from $\mathcal{R}$; it involves a further randomization, depending on the time spent in another set S, which loosely speaking indicates the existence of a suitably large plus-phase droplet. With these ingredients one is able to obtain a result that resembles that in Theorem 1, i.e. there exists t_h so that, for all $t > 0$

$$\lim_{h \to 0} P \left(\frac{T_{\beta,h}}{t_h} > t \right) = e^{-t} \tag{21}$$

and (see (20))

$$\lim_{h \to 0} h \log(t_h) = 3\Delta_c.$$

Notice that in this restricted setup the "escape time" is much larger than that suggested by the analysis of the infinite dynamics. It is not hard to understand why this should be true; in infinite (or very large) volume, a plus-phase droplet can appear anywhere and grow, reaching the origin in a shorter time. (See [30, 54, 55] for more details.)

Acknowledgements M. E. Vares acknowledges support of CNPq (grant 305075/2016-0) and FAPERJ (grant E-26/203.048/2016).

References

1. S. M. Allen and J. W. Cahn: A microscopic theory for antiphase boundary motion and its application to antiphase domain coarsening. Acta Metallurgica **27**, 1085–1095 (1979)
2. J. Barrera, O. Bertoncini, R. Fernández: Abrupt convergence and escape behavior for birth and death chains. J. Stat. Phys. **137** (4), 595–623 (2009)
3. F. Barret: Sharp asymptotics of metastable transition times for one dimensional spdes. Ann. Inst. H. Poincaré Probab. Statist. **51** (1), 129–166 (2015)
4. D. J. Barsky, G. Grimmett, C. M. Newman: Percolation in half-spaces: equality of critical probabilities and continuity of the percolation probability. Probab. Theory Relat. Fields **90** (1), 111–148 (1991)
5. N. Berger, C. Borgs, J. T. Chayes, A. Saberi: Asymptotic behavior and distributional limits of preferential attachment graphs. Ann. Probab. **42**, 1–40 (2014)
6. C. Bezuidenhout, G. Grimmett: The critical contact process dies out. Ann. Probab.**18** (4), 1462–1482 (1990)
7. A. Bianchi, A. Gaudillière: Metastable states, quasi-stationary distributions and soft measures. Stochastic Process. Appl. **126** (6), 1622–1680 (2016)
8. A. Bianchi, A. Gaudillière, P. Milanesi: On soft capacities, quasi-stationary distributions and the pathwise approach to metastability. arXiv:1807.11233
9. A. Bovier, M. Eckhoff, V. Gayrard, and M. Klein: Metastability in reversible diffusion processes I: Sharp asymptotics for capacities and exit times. J. Eur. Math. Soc. **6**, 399–424 (2004)
10. A. Bovier, F. den Hollander: Metastability: A potential theoretic approach. Springer (2015)
11. S. Brassesco: Some results on small random perturbations of an infinite dimensional dynamical system. Stoch. Proc. Appl. **38**, 33–53 (1991)

12. S. Brassesco, E. Presutti, V. Sidoravicius, M. E. Vares: Ergodicity of a Glauber+Kawasaki process with metastable states. Markov Proc. Relat. Fields **6** (2), 181–203 (2000)
13. V. H. Can. Metastability for the contact process on the preferential attachment graph. Internet Math. 45pp. (2017)
14. N. Chafee and E. F. Infante: Bifurcation and stability for a nonlinear parabolic partial differential equation. Bull. Am. Math. Soc.**80**, 49–52 (1974)
15. S. Chatterjee, R. Durrett: Contact process on random graphs with degree power law distribution have critical value zero. Ann. Probab. **37**, 2332–2356 (2009)
16. J. W. Chen: The contact process on a finite system in higher dimensions, Chinese J. Contemp. Math. **15** 13–20 (1994)
17. M. Cramston, T. Mountford, J.-C. Mourrat, D. Valesin: The contact process on finite homogeneous trees revisited. Alea **11** (1), 385–408 (2014)
18. M. Cassandro, A. Galves, E. Olivieri, M. E. Vares: Metastable behaviour of stochastic dynamics: a pathwise approach. J. Stat. Phys. **35**, 603–634 (1984)
19. A. De Masi, P. A. Ferrari, and J. L. Lebowitz: Reaction-diffusion equations for interacting particle systems. J. Stat. Phys. **44**, 589–644 (1986)
20. A. Debussche, M. Hoegele, and P. Imkeller: Asymptotic first exit times of the Chafee-Infante equation with small heavy-tailed Lévy noise. Electron. Commun. Probab. **16**, 213–225 (2011)
21. A. Debussche, M. Högele, and P. Imkeller: The Dynamics of Nonlinear Reaction-Diffusion Equations with Small Lévy Noise, Lecture Notes in Mathematics **2085**, Springer (2013)
22. R. Durrett: On the growth of one dimensional contact process. Ann. Probab. **8** (5), 890–907 (1980)
23. R. Durrett: Random Graph Dyamics. Cambridge Univ. Press, Cambridge (2007)
24. R. Durrett, X-F. Liu: The contact process on a finite set. Ann. Probab. **16** (3), 1158–1173 (1988)
25. R. Durrett, R. H. Schonmann: The contact process on a finite set II. Ann.Probab. **16** (4), 1570–1583 (1988)
26. J. Farfan, C. Landim, K. Tsunoda: Static large deviations for a reaction-diffusion model. arXiv:1606.07227 (2016)
27. W. G. Faris and G. Jona-Lasinio: Large fluctuations for a nonlinear heat equation with noise. J. Phys. A **15**, 3025–3055 (1982)
28. M. I. Freidlin and A. D. Wentzell: *Random Perturbations of Dynamical Systems*. Grundlehren der mathematischen Wissenschaften. Springer, Berlin- Heidelberg (1998)
29. A. Galves, E. Olivieri, and M. E. Vares: Metastability for a class of dynamical systems subject to small random perturbations. Ann. Probab. **15**, 1288–1305 (1987)
30. A. Gaudillière, P. Milanesi, M. E. Vares. Asymptotic exponential law for the transition time to equilibrium of the metastable kinetic Ising model with vanishing magnetic field. arXiv:1809.07044
31. T. Harris: Contact interactions on a lattice. Ann. Probab. **2**, 969–988 (1974)
32. D. Henry: *Geometric theory of semilinear parabolic equations*. Lecture Notes in Mathematics **840**, Berlin-Heidelberg-New York: Springer-Verlag., (1981)
33. A. Hinojosa: Exit time for a reaction diffusion model. Markov Processes and Related Filelds **10** (4), 705–744 (2005)
34. M. Högele and I. Pavlyukevich: Metastability in a class of hyperbolic dynamical systems perturbed by heavy-tailed Lévy type noise. Stochastics and Dynamics **15**(3) (2015)
35. H. A. Kramers: Brownian motion in a field of force and the diffusion model of chemical reactions. Physica **7** (4), 284–304 (1940)
36. D. A. Levin, M. Luczak, and Y. Peres: Glauber dynamics for the Mean-field Ising Model: cut-off, critical power law, and metastability. Probab. Theory Rel. Fields **146**, 233–265 (2010)
37. T. M. Liggett: Interacting Particle Systems. Springer, New York (1985)
38. T. M. Liggett: Stochastic Interacting Systems: Contact, Voter and Exclusion Processes. Springer, Berlin (1999)
39. F. Martinelli, E. Olivieri, and E. Scoppola: Small random perturbations of finite and infinite-dimensional dynamical systems: Unpredictability of exit times. Journal of Statistical Physics **55**, 477–504 (1989)

40. T. S. Mountford: A metastable result for the finite multidimensional contact process. Can. Math. Bull. **36** (2), 216–226 (1993)

41. T. S. Mountford: Existence of a constant for finite system extinction. J. Stat. Phys. **96** (5/6), 1331–1341 (1999)

42. T. Mountford, J.-C. Mourrat, D. Valesin, Q. Yao: Exponential extinction time of the contact process on finite graphs. Stoch. Proc. Appl. **216**, 1974–2013 (2016)

43. T. Mountford, D. Valesin, Q. Yao: Metastable densities for the contact process on power law random graphs. Electron. J. Probab. **18**, 1–36 (2013)

44. J.-C. Mourrat, D. Valesin: Phase transition of the contact process on random regular graphs. Electron. J. Probab.**21**, 1–17 (2016)

45. E. J. Neves, R. H. Schonmann: Critical droplets and metastability for a Glauber dynamics at very low temperatures. Commun. Math. Phys. **137**, 209–230 (1991)

46. E. J. Neves, R. H. Schonmann: Behaviour of droplets for a class of Glauber dynamics at very low temperatures. Probab. Theory Relat. Fields **91**, 331–354 (1992)

47. E. Olivieri, M. E. Vares: Large deviations and metastability. Cambridge University Press (2005)

48. R. Pemantle: The contact process on trees. Ann. Probab. **20**, 2089–2116 (1992)

49. O. Penrose, J. L. Lebowitz: Rigorous treatment of metastable states in the van der Waals-Maxwell Theory. J. Stat. Phys. **3**, 211–241 (1971)

50. M. Salzano: The contact process on graphs. PhD thesis, UCLA, (2000). (Reprinted *Publicações Matemáticas*. IMPA, 2003.)

51. M. Salzano, R. Schonmann: A new proof that for the contact process on homogeneous trees local survival implies complete convergence. Ann. Probab. **26**, 1251–1258 (1998)

52. R. H. Schonmann: Metastability for the contact process. J. Stat. Phys. **41** (3/4), 445–484 (1985)

53. R. H. Schonmann: The pattern of escape from metastability of a stochastic Ising model. Commun. Math. Phys. **147**, 231–240 (1992)

54. R. H. Schonmann: Theorems and conjectures on the droplet driven relaxation of stochastic Ising model. In *Probability and Phase Transition*, ed. G. Grimmett. NATO ASI Series. Dordrecht, Kluwer, 265–301 (1994)

55. R. H. Schonmann, S. Shlosman: Wulff droplets and the metastable relaxation of kinetic Ising models. Commun. Math. Phys.**194** (2), 389–462 (1998)

56. A. Simonis: Metastability of the d-dimensional contact process. J. Stat. Phys. **83** (5/6), 1225–1239 (1996)

57. M. Stacey: The existence of an intermediate phase for the contact process on tress. Ann. Probab. **24**, 1711–1726 (1996)

58. Y. Zhang: The complete convergence theorem of the contact process on trees. Ann. Probab. **24**, 1408–1443 (1996)

How Mathematics Is Changing the World

Maria J. Esteban

Abstract Mathematics was always, and is now more than ever, a key technology for innovation. Not many people would understand this sentence, because there is a wide-spread belief that Mathematics is not really useful except for teaching purposes. But actually the above statement can be made without any problem, because Mathematics is playing an increasing role in the development of new technologies, and its influence is only going to increase in the future. In the talk given in the WMWM meeting I tried to "prove" the above statements with examples.

1 The Message Behind the Talk Given at the WM[2]

There are three keywords to explain why Mathematics is so important for innovation. And they are: **Modeling, Simulation and Optimization (MSO)**.

Modeling means that using mathematical language, mathematical functions, mathematical equations, it is possible to describe natural (physical, mechanical, chemical, biological, . . .) phenomena and their evolution. Once a phenomenon is described by a given model, this model can be studied in order to understand how it will develop in different conditions and under different influences. A large class of possible situations can be taken into consideration. In each of those, the phenomenon or process can be studied mathematically, but more often using discretization and computational means. This is what is called **Simulation**. One can simulate a given process, in a given situation, with the help of a model and a computer (or many computers). Once this is done, one can also optimize over the conditions under which the process takes place. That is, one can choose a criterion, or several ones, to select the optimal result. This can consist in optimizing time, cost, energy, etc. This last part of the full process is what is naturally called **Optimization**.

M. J. Esteban (✉)
CEREMADE (CNRS UMR n° 7534), Université Paris-Dauphine, PSL Research University, Paris 16, France
e-mail: esteban@ceremade.dauphine.fr

© The Association for Women in Mathematics and the Author(s) 2019
C. Araujo et al. (eds.), *World Women in Mathematics 2018*, Association for Women in Mathematics Series 20, https://doi.org/10.1007/978-3-030-21170-7_4

Let us discuss **Modeling** in more detail. What is it? First, a given phenomenon or process is detailed in a quantitative way, in the form of equations, or functions, or mathematics objects and their relations. These equations can describe the behavior of fluids, solids, waves, sound, heat, the deformation of a solid, the combustion of an engine, or the evolution of an epidemic of some illness in a given medium. This is what modeling can do for natural phenomena, to quantify the relations between the different aspects of what one wants to describe: force, speed, deformation, intensity, mass, energy, etc. But modeling can be also used to describe a process, like for instance how to clean a blurred image or an old movie. Or how to organize a network's functioning in an optimal way. Or how to code information on internet securely.

In the past, many physical and mechanical processes were dealt with "by hand" and by building and using prototypes. Modeling allows us to overcome that costly and lengthy process and becomes a way to accelerate the birth and development of new technologies. In some cases it is not only a case of reducing the cost, or the time needed to build a new object, a plane, a machine, a car... but in many cases prototyping is more and more out of reach, or too dangerous, while modeling is always possible. Other possible uses of modeling can help, for instance, to describe the organization of a large set of objects or people, like the planes and the personnel of a big transportation company. This is what is called logistics.

Modeling can be done in a deterministic way, using equations or functions relating different variables, or it can be done using probabilistic or statistical functions and concepts when there is some randomness in the underlying process. Or using other concepts coming from various fields of Mathematics. The situation will ask for one or another, and often various approaches are possible and different groups will use different approaches to address the same question. When modeling is done to describe a problem in a scientific field which is not Mathematics, like Physics, or Biology for instance, this work is often done by specialists in that field, or by them and mathematicians together. This is natural since in order to depict a phenomenon well, one has to understand the underlying processes, and for that a specialist is often needed. Also, the phenomena or processes one wants to model are frequently too complex to be treated, and so one has to choose how to simplify the modeling in a way that it still makes sense. This means that the important part of the model has to be kept in it, and some other parts, the less relevant details, those that will not play an important role in the result, at least not in a significant way, maybe be forgotten, at least for some time. This again has to be decided by specialists who know what is the most important part that has to be kept, and what can be neglected in the first stage.

Note that modeling can involve continuous or discrete mathematics, and even if in the past the natural mathematical fields most used in modeling were the so-called applied mathematics ones, like differential equations, probability and statistics, numerical analysis, etc, nowadays practically all fields of mathematics can help to deal with applications, like algebra, geometry, topology, number theory, etc.

Modeling can be used to solve problems in all kinds of societal and industrial fields, in logistics, to organize the material and personnel organization of big transportation companies; in manufacturing, to help designing machines and parts of machines needing a sophisticated design; to help in building new efficient engines, airplanes, cars, smartphones; imagining smart cities; supervising and controlling pollution; to help in finding optimal therapies for cancer and other illnesses; or designing the optimal shape of a bypass; in decision making using so-called operations research theory, etc.

Then comes **Simulation**. What is this about? The models that one studies in most situations are impossible to be solved in an analytical or exact way. Mathematics can be used to prove that the problem under study has solutions or not, and if yes, what kind of properties they enjoy. But if one wants to know the solutions more concretely, and this is often of the utmost importance in applications, computers have to be used to calculate approximate solutions. This means that the model has to be discretized, approximating an infinite number of points or dimensions by a finite number of them, and then trying to solve the problem in that discrete set-up. This can be again done in many different ways, and mathematicians are endlessly improving the properties of the algorithms they use to solve a particular equation or system of equations, or model, this being done in order to obtain an ever better approximation. Once one has the discrete model, it can be implemented and solved with the help of computers. This is the meaning of simulation for a given problem, to study it in an approximate way with the help of computers and so "see" the solution, or the evolution of functions, etc, in a concrete way. The solutions of course will not be exact, but, if the discretization and the algorithm used are good, they will give a very good idea about the exact solutions that cannot be known. And in many cases one can even measure how far from each other the exact and the approximate solutions are.

The third methodology which helps to make Mathematics so useful for innovation, industry and for the design and treatment of societal activities is **Optimization**. And this goes together with the two previous ones. In the design of a model or an algorithm many choices are made, about the constraints, about the relevant parameters which characterize the situation in which the experiment or activity takes place, etc. How to make those choices in an optimal manner is not known a priori. Optimization means that one chooses, in the models, in the discretization, in the algorithms, the values of the parameters or of the constraints that yield the best result. Best in which sense? This will be based on some well-chosen criteria: one can want to optimize energy, time, cost, quantity of material, money, etc. And how is this achieved? The simulation can be done for different sets of values of the parameters and then the results of the different computations compared. Then, the best set of parameters is chosen in the end. But one can also use optimization techniques that allow us to know a priori how to proceed to get optimal results.

These three aspects of how and why Mathematics is so important for applications in the real world are the basis of the whole construction. But they can be complemented with other branches of Science and computational methods involving for instance *Artificial Intelligence* (AI) aspects, data analysis, statistical criteria, etc.

We will conclude the paper by considering these branches. Apart from this aspect, it is good to mention that many people, and not only mathematicians, can model, simulate and optimize. Or use codes designed to do this. What a mathematician can add is to make all the above with a certainty of obtaining good results, or at least being able to measure the errors made in a simulation, or showing how to control the instabilities that can arise from tiny variations in the data. Mathematicians prove theorems, and not only of existence, stability, etc. They can also prove theorems about the discretizations and algorithms that they devise in order to find approximate solutions for a given model or problem. And those theorems can provide the users with vital information about error estimates, speed of convergence of an algorithm, stability, robustness, reliability, etc. Mathematicians can thus provide results which are robust and guaranteed. And which go together with measures of the error or the reliability. This is of course a big plus for a company, which wants to be sure that their products will be good and competitive. Or their processes optimal and efficient.

Nowadays the use of *Modeling, Simulation, Optimization* (MSO) together with machine learning and AI is becoming a must. And a new and very interesting concept, very much liked by advanced industry, is that of *digital twins*. A digital twin is a digital model for physical assets, processes and systems that can be used for various purposes. It integrates MSO with artificial intelligence, machine learning and software analytics with data to create living digital simulation models that update and change continuously. Again mathematicians can, and should, play an important role in the creation and maintenance of digital twins. And they have to do it with engineers and other scientists, experts in the fields concerned with the model. The integration of different kinds of expertise is a guarantee to enhance the success of the twin's results.

Many people use to say that Mathematics is the language of Science. And the final report of the European *Forward Look for Mathematics in Industry* [1] said that Mathematics is also the language of innovation. The above statements about the possibilities of Mathematics and mathematicians to help solve real problems and to help companies produce better technologies, and do it in a more competitive and efficient way, is not just blah blah: they have recently been made more precise. Indeed, there have been several independent studies proving and quantifying the economic impact of Mathematics on the economy of three European countries. The impact studies of Mathematics on the British [2], Dutch [3] and French [4] economies and societies have shown incredible numbers, thus proving that investing in Mathematics is really worth it! Let us conclude this text by pointing out that more data and information relevant to the subject of this talk, and many more examples, can be found in various documents available in the *Reports* section of the EU-MATHS-IN website: http://www.eu-maths-in.eu/EUMATHSIN/reports/

References

1. Mathematics and Industry - Report. http://www.eu-maths-in.eu/EUMATHSIN/wp-content/uploads/2018/01/FLMI-Report-final.pdf
2. Measuring the Economic Benefits of Mathematical Science Research in the UK. http://www.eu-maths-in.eu/EUMATHSIN/wp-content/uploads/2016/02/2012_UK_EconomicBenefitsOfMathematicalScienceResearch.pdf
3. Mathematical sciences and their value for the Dutch economy. http://www.eu-maths-in.eu/EUMATHSIN/wp-content/uploads/2016/02/2014_Netherlands_MathematicalSciencesValueToEconomy.pdf
4. Etude de l'impact socio-économique des Mathématiques en France. http://www.eu-maths-in.eu/EUMATHSIN/wp-content/uploads/2016/02/2015-France-SocioEconomic_Impact_of_Mathematics.pdf

Part II
The Gender Gap in Mathematical and Natural Sciences from a Historical Perspective

A Data Analysis of Women's Trails Among ICM Speakers

Helena Mihaljević and Marie-Françoise Roy

Abstract The International Congress of Mathematicians (ICM), inaugurated in 1897, is the greatest effort of the mathematical community to strengthen international communication and connections across all mathematical fields. Meetings of the ICM have historically hosted some of the most prominent mathematicians of their time. Receiving an invitation to present a talk at an ICM signals the high international reputation of the recipient, and is akin to entering a 'hall of fame for mathematics'. Women mathematicians attended the ICMs from the start. With the invitation of Laura Pisati to present a lecture in 1908 in Rome and the plenary talk of Emmy Noether in 1932 in Zurich, they entered the grand international stage of their field. At the congress in 2014 in Seoul, Maryam Mirzakhani became the first woman to be awarded the Fields Medal, the most prestigious award in mathematics. In this article, we dive into assorted data sources to follow the footprints of women among the ICM invited speakers, analyzing their demographics and topic distributions, and providing glimpses into their diverse biographies.

1 The Hall of Fame for Mathematics

Ever since its inaugural gathering in 1897, the International Congress of Mathematicians (ICM) has signified the greatest effort of the mathematical community to establish international communication and connection across all mathematical topics. Throughout their history, the congresses have hosted some of the most prominent mathematicians of their time. Needless to say, receiving an invitation to present a talk at an ICM is a matter of high international reputation, often compared

H. Mihaljević
Hochschule für Technik und Wiertschaft Berlin, University of Applied Science, Wilhelminenhofstraße 75A, 12459 Berlin, Germany
e-mail: Helena.Mihaljevic@HTW-Berlin.de

M.-F. Roy (✉)
IRMAR (UMR CNRS 6625), Université de Rennes 1, Rennes Cedex, France
e-mail: marie-francoise.roy@univ-rennes1.fr

© The Association for Women in Mathematics and the Author(s) 2019
C. Araujo et al. (eds.), *World Women in Mathematics 2018*, Association for Women in Mathematics Series 20, https://doi.org/10.1007/978-3-030-21170-7_5

111

with the entrance into a 'hall of fame for mathematics'. In fact, it is no exaggeration to state that an ICM invitation is often treated like the reception of a major research award.

Women mathematicians attended the ICM from the start, not only as accompanying persons but also participating on their own, e.g as professional mathematicians. Nevertheless, female speakers remained very few. The share of women in selected congresses has been addressed in some previous works. Fulvia Furinghetti studied the presence and contribution of women to the discipline of mathematics education in the first half of the twentieth century using data from two scientific journals, the proceedings of the first International Congress on Mathematical Education (ICME) in 1969 and the didactics sections of the ICM proceedings until 1966 [5]. She describes the difficulties posed by differences in structure of the individual congresses, the layout of the congress proceedings, or inconclusive and incomplete data (e.g. regarding the distinction of 'accompanying persons'). She provides numbers of women among participants and contributors for ICMs until 1966 and gives insights into biographies of women pioneers. The essays [16] by Cora Sadosky and [3] by Bettye Anne Case and Anne M. Leggett from the collection 'Complexities. Women in Mathematics' address the participation of women lecturers since 1974, focussing mainly on the collective efforts of women in the 1970s and 1980s to overcome their persistent underrepresentation as invited congress speakers. Both pieces arrive at similar conclusions, namely, that the actions in the 1970s and 1980s have strongly contributed to the diversification of the congress, yielding a significantly higher chance for qualified women to be invited to speak.

While the mentioned research addresses the participation of women at individual congresses or throughout certain periods, to our knowledge there is no global exploratory analysis of the demographics of ICM speakers from its beginning until today. In this contribution, we thus investigate data on all invited ICM speakers from 1897 to 2018. Using various data sources, in particular the list of all invited speakers from Wikipedia, Wikidata pages of individual speakers, and the subdivision of congress speakers into sections from the International Mathematical Union (IMU), we are able to address the following questions regarding women's participation: How inclusive has the congress been throughout its history? What factors might have positively influenced the share of women? Are there noteworthy differences between women and men speakers regarding age, country of residence or research areas?

We start out by describing pioneer contributions of women. We then outline the development of women's participation in the congresses over time, elaborating some of the advances and setbacks. Finally, we investigate the distribution of women and men speakers by countries of citizenship and sections of the delivered talk.

2   Data Basis and Methods

In February 2018, we programmatically extracted all names and ICM dates from the Wikipedia website 'List of International Congresses of Mathematicians Plenary and Invited Speakers' [11], which resulted in a table containing 3,745 plenary and invited speakers from 1897 until 2014. Using the hyperlinks contained therein, we retrieved the gender, country of citizenship, date of birth and employer from Wikidata, a free, human- and machine-readable knowledge database that serves as a central storage for structured data of other Wikimedia projects, including Wikipedia.[1] We have found a Wikidata page for 82.6% of all listed speakers, and 77.5% of all unique individuals (various mathematicians gave multiple talks). A Wikidata page existed for almost all women, namely 92.5%. The coverage shows a certain trend: with exception of the large congresses in 1928 and 1932, the vast majority of speakers from the early congresses (usually 90% or higher) has a page in Wikidata, with decreasing trend over time.

For speakers invited to the ICM 2018 in Rio de Janeiro, we extracted their names, country of citizenship and the ICM sections of their talks from the official ICM-2018 website. We used Python package gender-guesser,[2] which has shown very reliable results in a recent benchmark on name-based gender inference [17], to infer the gender[3] of the speakers using their forenames when this information was missing. For speakers whose names are not highly correlated with only one gender (across different countries and languages), and for which gender-guesser hence did not produce a definite gender assignment, we filled this information manually, mainly based on field knowledge and Internet research.

The International Mathematical Union (IMU) provided us with a file containing speaker names, ICM date and place, and the name of the section of the corresponding talk (see [7] for a search interface within the official IMU website). Due to different name spellings in the datasets, we applied fuzzy string matching techniques to combine the data sources and add the sections to our original data set. For many speakers at the congress in 1950 the section was missing and hence needed to be filled manually.

In addition, we added the date of birth and country of citizenship for all women speakers in order to create a data basis which is as complete as possible and that can be used for information and teaching purposes beyond this analysis. For this purpose we have contacted those women in our list for whom information was still missing. We have made the list of all speakers available at [12]. We noted that there exist different countings of invited ICM speakers. For instance, the list of speakers provided by the IMU [7] contains around 400 speakers more than the list at Wikipedia [11], in particular for the congresses before 1950. This is mainly due to the change of terminology over time and the respective counting schemes. On the other hand, the list at Wikipedia contains speakers who were invited but did not attend. Our analyses are based on the Wikipedia list [11] which applies the post World War II terminology in which the one-hour speakers in the morning

---

[1] Every Wikipedia article is supposed to have a corresponding entry in Wikidata.

[2] https://pypi.org/project/gender-guesser/.

[3] For all authors we used a gender assignment provided by a third party (Wikidata or a web service) which, for our dataset, resulted in a binary schema.

sessions are called 'Plenary Speakers' and the usually more numerous speakers (in the afternoon sessions) whose talks are included in the ICM published proceedings are called 'Invited Speakers'" [11]. Usually, there were a lot more additional shorter contributions that were not always part of the congress proceedings. Moreover, the list of speakers from Wikipedia [11] does not reflect whether a speaker gave more than one talk at a given congress. This, in fact, was not so rare; for example at the congress in 1900 in Paris, Gösta Mittag-Leffler gave both a plenary talk and one in the *Analysis* section. In order to take into account such multiple contributions, we have expanded the data using the sections from the list supplied by the IMU [7].

3   Women Pioneers

The organizers of the congress in 1908 in Rome invited Laura Pisati, the first woman to present a paper. Not much is known about her personal and professional life, other than that Pisati was an active mathematics researcher and the author of internationally recognized publications. In zbmath.org, we find a book and three research articles listed in her author profile [19], two of them published in the influential *Rendiconti del Circolo Matematico di Palermo*, the journal of the Mathematical Circle of Palermo, of which she was a member. In 1905, she also became a member of the German Mathematical Society [6, p.12]. According to her membership information, Pisati was born in Ancona (date of birth not listed). She graduated in mathematics from the University of Rome in 1905 [15]. Since 1897 she had worked as a teacher at the Technical School 'Marianna Dionigi' in Roma (Scuola Tecnica 'Marianna Dionigi' di Roma), one of the the first secondary schools for girls in Rome. She was engaged to Giovanni Giorgi, an Italian physicist and electrical engineer. In 1900, Pisati had been entrusted with the supervision of his thesis in Mathematics [9]. Sadly, she died young on March 30 1908, only a few days before the 1908 congress in Rome and before her planned wedding to Giorgi. Her paper 'Saggio di una teoria sintetica delle funzioni di variabile complessa' was presented by a male colleague.

In the report on the sectional meetings of the congress [14], Laura Pisati appears as the only speaker with first and last name listed, showing the singularity of women's presence in this circle at that time. Interestingly, Giorgi himself was an invited speaker at three subsequent ICMs, in 1924 in Toronto, in 1928 in Bologna, and in 1932 in Zurich. He cited Pisati's work in his 1924 ICM contribution with the words "See also some very striking results given by LAURA PISATI in her paper Sulle operazioni funzionali non analitiche originate da integrali definiti. Rend. Cire. Mat. Palermo, Tomo XXV (1908) pp. 272–282." [4, p.45].

Four years later, in 1912, Hilda Hudson was the first woman to speak at an ICM with a paper she presented in the *Geometry* section. Hudson, a member of a family of distinguished mathematicians, worked mainly in the theory of Cremona transformations, on which she had published various articles. Between 1910 and

Hostinský, Dr B., Kr. Vinohrady, Moravska 40, Prague. *St John's College.*
Huard, Professor A., Lycée Henri IV, Paris. *King's College.*
Hubrecht, J. B., The Hill House, Coton.
 Hubrecht, Mme M.
Hudson, Professor W. H. H., 34, Birdhurst Road, Croydon. *St John's College.*
 Hudson, Miss. , *Newnham College.*
Huntington, Professor E. V., Harvard University, Cambridge (Mass.). 15, *Holland Street.*
 Huntington, Mrs.
Hunton, Professor S. W., Mount Allison University, New Brunswick, Canada. c/o
 A. Berry, Meadowside, Grantchester Meadows, Cambridge.
Hutchinson, A., Aysthorpe, Newton Road, Cambridge.
 M. C.

2

Fig. 1 Hilda Phoebe Hudson, the first woman who presented her work at an ICM, listed as an accompanying participant at the ICM 1912 in Cambridge

1913, she was an Associate Research Fellow at the Newnham College[4] [2]. As pointed out in [5], Hilda Hudson is listed in the Proceedings of the congress in 1912 as an accompanying person to her father, Prof. William Henry Hoar Hudson, showing how misleading the distinction between accompanying persons and 'real' participants was in that period (Fig. 1).

The 1932 congress in Zurich witnessed the first plenary talk by a woman, given by Emmy Noether, who spoke about hypercomplex systems in their relations with commutative algebra and number theory.[5] Her invitation certainly marked a milestone in the representation of women within the international mathematical community. Noether had already attended previous congresses. At the age of 26 she accompanied her father, Max Noether, who spoke at the congress in 1908 in Rome, where Pisati was supposed to present her work. Prior to her plenary lecture in 1932, Emmy Noether gave a talk at the congress in Bologna four years earlier. As the positive trend in the early years of the ICM did not persist, it was almost 60 years until Karen Uhlenbeck became the second woman to give a plenary talk at an ICM under the title 'Applications of non-linear analysis in topology' Fig. 2.

In 2014 at the ICM in Seoul, Maryam Mirzakhani was awarded the Fields Medal for "her outstanding contributions to the dynamics and geometry of Riemann surfaces and their moduli spaces".[6] She is the only woman among the 60 mathematicians who have received the Fields Medal, a prize conferred since 1936 to at most four mathematicians at each congress under the age of 40. Mirzakhani was diagnosed with breast cancer in 2013 and died on July 14, 2017, at the age of 40.

---

[4]Newnham College, founded in 1871, was the second women's college to be established in Cambridge. It acquired full university status in 1948, the year in which the first women were were officially admitted to the University.

[5]Original title of the talk in German: "Hyperkomplexe Systeme in ihren Beziehungen zur kommutativen Algebra und zur Zahlentheorie".

[6]ICM laudation, http://www.icm2014.org/en/awards/prizes/f4.

Fig. 2 Left: Emmy Noether (front) on a steamboat trip during ICM-1932. (ETH-Library, Zurich). Right: Karen Uhlenbeck in 1982, eight years before her plenary lecture in Kyoto (Oberwolfach Photo Collection)

4 The History in Numbers: Advances and Setbacks

Out of 4,120 invited contributions from 1897 to 2018, 202 were presented or authored by women, which amounts to only 5% of the total. Women's participation over time, however, did not grow steadily but, instead, shows multiple trends. As presented in Fig. 3, a comparatively large number of women presented their research at the congresses in 1928 at Bologna and in 1932 at Zurich. This reflects the overall progressive societal and political spirit of the 1920s, which had also enhanced the situation of women in science. In fact, the ICM in 1932 marks a pinnacle in the history of ICMs regarding the role of women. Emmy Noether gave the first plenary lecture by a woman; various women's colleges and organizations of university women sent delegates, among them the Bedford College for Women (London), Hunter College (New York), the International Federation of University Women, and the American Association of University Women.

ICMs were always affected by global political events. The first substantial tension occurred in the aftermath of World War I, as mathematicians from Germany were excluded during the ICMs in 1920 and 1924. The sole choice of Strasbourg as the location for the congress in 1920 was a political statement in itself. Already in the 1920s, Italy, which experienced a golden era in both pure and applied mathematics at the turn of the nineteenth century, showed the first signs of a deep crisis caused by a spreading fascism [8]. The last congress before World War II that took place in 1936 in Oslo was signified by different political agendas, in particular by the German strategy to present 'Aryan mathematics'. Italian mathematicians boycotted the congress, Soviet mathematicians were denied the travel permission by their political authorities. That only few women were invited to the congress in Oslo seems not too surprising given the political situation at that time—the spread of Fascism through Europe, persecution of Jewish mathematicians, and the worldwide

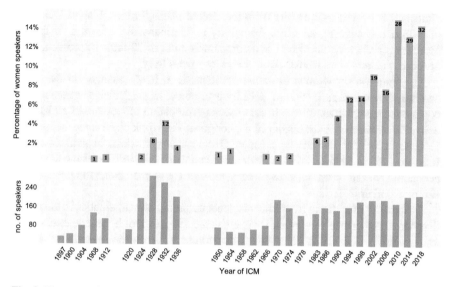

Fig. 3 Upper panel: Bar height shows the percentage of women speakers per ICM, the numbers inside correspond to the total numbers of invited women. Lower panel: total number of speakers per year

economic depression—which in some aspects affected women mathematicians on a larger scale. When the first congress after World War II took place in 1950 in Cambridge (USA), the only woman lecturer was Mary Cartwright, at that time Mistress of the Girton College.[7]

It took 60 years to reach a share of women among ICM speakers comparable to that in 1932. Among the manifold reasons for this situation are undeniably the impact of some historical and political developments. The aftermath of World War II was characterized by a rollback in society as a whole. The 1950s experienced a return to conservative gender roles, in which women were expected to take care of the domestic sphere, leaving the work places to the men who were coming back from the battlefields. These conceptions had impact on university education as well. During the conservative post-war era in Germany, for instance, the share of female students decreased significantly, and there was general agreement that men should take precedence in accessing the limited study places. However, some countries managed to overcome some of these barriers in women's university education and research faster than others. Partially, these general trends are also reflected in country-based differences regarding the presence of women speakers at postwar ICMs: in the 11 congresses between 1950 and 1990, of the 24 talks given by women,

---

[7]Girton College was the first women's college to be established in Cambridge. It began in Hitchen (about 24 miles from Cambridge) in 1869 before moving to Girton in 1873, when it acquired the name Girton College. Like Newnham, it obtained full college status in 1948.

almost all delivered by speakers from the United States, France, United Kingdom, or Russia but none by speakers from Italy or Germany. By contrast, in the ten congresses before World War II of a comparable total of 27 talks by women, three of those speakers were from Germany and four from Italy.

The situation for women as active participants in ICMs changed in the 1990s and has shown a certain level of stability ever since. In particular, the recent three congresses have witnessed a hitherto unseen participation of women: of all lectures delivered by women in the history of the congress, 80% took place since the meeting in 1990 in Tokyo. The drastic change affects plenary sessions in particular: The ICMs in 2002 in Beijing, 2010 in Hyderabad and 2018 in Rio de Janeiro collectively accounted for ten of the total 18 plenary lectures by women since the premiere by Emmy Noether in 1932.

Despite the overall progress towards gender equality in mathematics in the recent decades, the increase of women speakers since 1990 cannot be interpreted simply as a positive side effect of a global development. A closer look at the events during the congresses shows that the increased invitation of women speakers is also, and maybe above all, the result of interventions by groups and individuals at various levels. As described in [3, 16], since 1974, organizations of women such as the Association for Women in Mathematics (AWM) have set up events during the congresses, often sparking discussions on what was often perceived as a systematic omission of women as invited speakers. At various congresses in the 1970s and 1980s resolutions were passed with the aim to increase the number of lectures by women. At ICM-1974, concerns about the small number of women speakers were raised during a discussion by the AWM. At the next ICM in 1978, a public protest initiated by AWM members resulted in a widely supported resolution to improve the situation of women in the future. Four women were invited to the congress in Warsaw in 1983, but there were no protests or reminders to keep improving the situation. It is probably no coincidence that in the program announcement of ICM-1986, not a single woman was listed in traditional mathematics research areas, suggesting that, as formulated by Sadosky in [16], "when there are no reminders about women mathematicians, colleagues tend not to remember us". The program of the ICM-1986 was changed on short notice, again through intervention, by presenting 25 qualified women candidates to the Executive Committee. The informal panel discussion organized by AWM on the situation of women in mathematics that took place during the congress in 1986 was at the origin of the constitution of the European Women in Mathematics (EWM).

The engagement of Mary Ellen Rudin in her role as the head of the U.S. delegates at the IMU General Assembly in 1986 is an illustrative example of what can change when individuals in prominent positions pursue this topic. The president of the ICM-1990 in Kyoto explicitly stated that the committees have followed Rudin's recommendation that subfields of mathematics, women, and mathematicians in small countries should not be overlooked [18].

Since 2010, specific satellite meetings of the congresses have been organized with the goal of highlighting the contributions and achievements, but also to address concerns of women mathematicians: the International Conference of Women

Mathematicians in 2010 in Hyderabad and the International Congress for Women Mathematicians in 2014 in Seoul. In Rio de Janeiro, the World Meeting for Women in Mathematics (WM^2) was set up as a satellite meeting combined with a panel discussion on the gender gap in the mathematical and natural sciences, and integrated into the ICM-2018 program and the ICM proceedings.

5 Any Difference?

Within the group of ICM speakers, men and women do show some differences regarding certain demographic aspects. For instance, both women and men speakers were around 44 years old when invited to give a lecture. However, before ICM-1950, women speakers were on average 36 years old, 9 years younger than their male colleagues. Since 1950, their average age has surpassed men's by almost 5 years.

We have focused on two particular aspects: the country of citizenship and the sections in which the speakers presented their research. The country of citizenship is interesting demographic information for ICMs, in particular due to their regional focus. The mathematical research fields, on the other hand, are known to show high variance in the share of women [13].

5.1 Distribution by Countries

We have collected the country of citizenship for 3,038 out of 3,987 speakers, mostly through their Wikidata pages. Further, we have undertaken additional manual efforts in order to collect missing information for all women speakers by using their websites, personal contacts or contacting them by e-mail.

Various countries listed in Wikidata do not exist anymore. We have thus replaced their names with those of today's states, e.g. Second Polish Republic with its successor state Poland or the Weimar Republic with Germany. When such a replacement is not possible, in particular for states that have disintegrated over the course of time such as Kingdom of Yugoslavia, Czechoslovak Socialist Republic or Austria-Hungary, we have inspected the demographics of the corresponding speakers to assign the closest state existing today.

Furthermore, for speakers listed with more than one country of citizenship we have weighted each of them by inverse frequency, e.g. for a speaker with citizenships of Germany and the United States, each would be counted as one half.[8] This

---

[8]The Wikidata entry of some speakers shows quite a few different citizenships, e.g. Đuro Kurepa, a plenary speaker in 1954 and 1958, has had 5 different citizenships according to his Wikidata entry: Socialist Federal Republic of Yugoslavia; Kingdom of Yugoslavia; Kingdom of Serbs, Croatians and Slovenes; Austria-Hungary; Federal Republic of Yugoslavia.

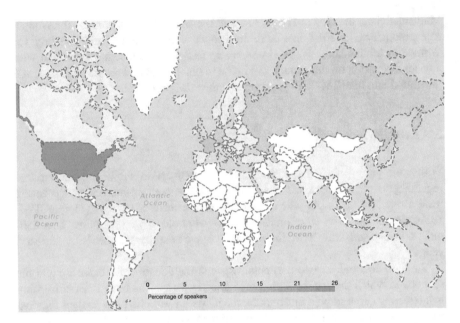

Fig. 4 Geographical distribution of all women speakers according to their country of residence

procedure was necessary since we do not have information on the time period in which a citizenship was valid. This aggregation thus presumably shows a certain bias for countries which are known to have attracted mathematicians, in particular the United States [1].

The map in Fig. 4 shows the proportion of countries as countries of citizenship among all women speakers. The overall distribution of geographical origins is, as expected, quite skewed: The six most frequent countries of citizenship among all speakers—United States (24.6%), France (13.2%), Germany (9.5%), Russia (8.5%), United Kingdom (7.4%) and Italy (7.2%)—comprise more than 70% of all. The evaluation for women yields a picture similar to the overall trend: almost the same countries appear under the top six, comprising more than 72% of all talks by women: United States (28%) and France (18.3%) are the most frequent countries, followed by Germany (8.3%) and United Kingdom (6.9%). Russia and Italy are less strong than in the distribution of all ICM speakers, with 5.5% and 4.6%, respectively. Italy, in fact, is not among the top six countries for women speakers, ranking at position seven, slightly behind Israel with a share of 5% among female speakers. Israel clearly marks the most notable difference in the comparison of origins, with an overall share of less than 2% compared to 5% among women speakers.

Almost all congresses show a certain regional focus, manifested in the composition of the organizing committees as well as the origin of the invited speakers (see Table 1 in Appendix). For instance, more than 32% of all speakers at the ICM-1900 in Paris were French; 45% of all speakers at the congress in Heidelberg in

1904 were German; United Kingdom was the country of citizenship for more than 26% of all speakers in 1912 in Cambridge (UK). The focus on 'local' speakers, evident in subsequent congresses as well, is especially pronounced for congresses that took place in countries with overall strong representation such as Russia at ICM-1966 in Moscow or the United States at ICM-1986 in Berkeley. Nevertheless, many other countries managed to foster the representation of their scientists. Brasil, for instance, accounts for 6% of all speakers at the latest congress in Rio de Janeiro. A look at the origins of women shows a rather inconclusive picture. A similar trend is expressed in a more drastic way: when a congress took place in a country with an overall strong representation, often all invited women originated from the host country. At most of these congresses, however, this corresponded to only one or two invited women. Nevertheless, at some congresses with a stronger representation of women such as the ICM in 1928 in Bologna or in 1986 in Berkeley, the host country reached a share of 40 to 50% among the invited women. On the other hand, there are 12 ICMs to which women were invited but none originated from the host country.

5.2   Topics Not Balanced

The individual character of the congresses is reflected by the diversity of names chosen for the themed sections, summing up to more than 180 different titles. While some section names such as *Numerical Methods, Numerical Mathematics* or *Numerical Methods and Computing* obviously belong together, this is much less the case with topics such as *History of Mathematics, Logic and Foundations* and *Mathematics Education*: at various ICMs, two or even all three of them had been combined together into one section, making it impossible to study these topics individually without intensive manual work. The division into few, rather broad sections was typical for the early ICMs. The ICM-1928 in Bologna, for instance, combined in the first section talks on *Algebra, Arithmetics and Analysis*, and had only one additional section on pure mathematics, mainly for talks in *Geometry*. *Elementary Mathematics, Didactical Questions*, and *Mathematical Logic* were grouped into one section, and *Philosophy and History of Mathematics* into another one. On the contrary, recent ICMs feature more than 20 themed sections, providing a better granularity to analyze the share of women speakers by their field of research.

The data by the IMU [7] shows certain inconsistencies, in particular for some early ICMs. For instance, Section 1 at ICM-1928 was named *Analysis* instead of *Arithmetics, Algebra, Analysis*. Furthermore, 15 talks from 1966 were assigned to the section 1/2 hr. report. The distinction between 1 hour and 1/2 hour is basically the distinction between plenary talks and section talks, and while most of the 1/2 hour talks were assigned to a themed section, these 15 remained for unknown reasons.

The partially unclean data for ICMs before 1970, and the fact that except for the congresses in 1928 and 1932 the presence of women before the 1980s is highly scattered, have prompted us to restrict our description of women's participation by sections to the period 1970–2018.

We have grouped the section names manually into the following eight groups (number of talks per section since ICM-1970 in parentheses): *Logic, Foundations, Philosophy, History and Education* (155); *Applied Mathematics, Applications of Mathematics, Mathematical Physics* (327); *Probability, Mathematical Statistics, Economics* (129); *Analysis, ODEs, PDEs, Dynamical Systems* (519); *Algebra and Number Theory* (271); *Theoretical Computer Science* (78); *Geometry and Topology* (515); *Combinatorics* (80). In addition, there were a total of 237 *Plenary* talks since 1970. There are two further categories of talks: the *ICM Abel Lecture* given to the winner of the Abel Prize (2 talks, both by men) and the *ICM Emmy Noether Lecture*[9] (6 talks by women). We have omitted both of them from the plot in Fig. 5.

As shown in Fig. 5, sections concerning *Algebra and Number Theory* have the least proportion of talks by women (<5%), closely followed by sections dealing with *Probability, Mathematical Statistics and Economics* (<6%). The two largest groups of sections, *Analysis, ODEs, PDEs, Dynamical Systems*, and *Geometry and Topology*, which together comprise more than 1,000 talks since 1970, have both less than 7% of women speakers on average. On the positive end we see two rather small section groups, *Logic, Foundations, Philosophy, History, Education* (>16%) and *Theoretical Computer Science* (>14%). The *Plenary* section, which is supposed to contain the most prominent congress talks, contains 8.2% talks by women and is hence slightly above the average of 7.3% since 1970.

Further investigation needs to be carried out to understand the unequal distribution among topics. In any case, there is no conclusive correlation between this distribution and the representation of women authors in the respective fields. The distribution of authorships and authors across classes of the Mathematics Subject Classification (MSC) 2010 has been carried out in [13] based on data from zbmath.org, one of the two main services for bibliographic information in Mathematics. Fig. 10 in [13] shows the amount of women authors as a heatmap across MSC classes. It shows, for instance, that an aboveaverage number of women publish in the field of Statistics and their share in Probability theory is close to the overall average. At the same time, the group of ICM sections related to *Statistics, Probability and Economics* is, with less than 6%, almost at the very bottom of the scale. Likewise, women authors are very well represented in most MSC classes related to *Analysis, PDEs, ODEs and Dynamical Systems*, in particular in relatively large fields like PDEs and ODEs. However, the respective group of ICM sections, while being the largest in terms of the total number of talks, has very few talks by women.

---

[9]The Emmy Noether Lecture honors women who have made fundamental and sustained contributions to the mathematical sciences. Since 2006, this lecture is a permanent ICM feature, since 2014, a special commemorative plaquette is conferred to every ICM Emmy Noether Lecturer.

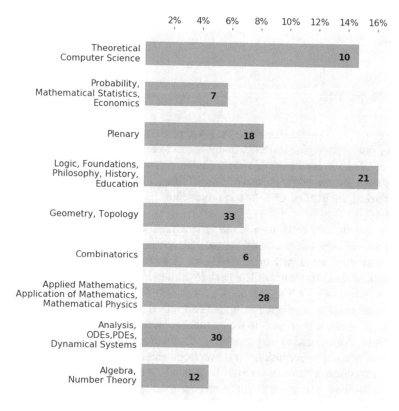

Fig. 5 Percentage of talks by women per group of sections since 1970. The numbers inside the bars correspond to the total number of talks by women

In [16], Cora Sadosky notes that the distribution of women across non-plenary sections suggests an "invisible quota system", leading to at most one woman per congress and section. She continues saying that "it seemed as if the selection panels, although aware enough to consider women candidates, felt that they had fulfilled their duty when the first one accepted." The first exceptions to this pattern occur in the section *Mathematical Aspects of Computer Science* featuring two talks by women in 1990 (Lenore Blum and Shafi Goldwasser) and in 1998 (Joan Feigenbaum and Toniann Pitassi). Also, the section related to *Teaching and Popularization of Mathematics* often contained more than one talk by a woman. In fact, most topics featured two talks by a woman at some congresses. In some rather exceptional cases, even more than two women were invited, such as in 2014 in *Combinatorics* or *Mathematics in Science and Technology*, or in 2018 in the *Geometry* section. Sections with an approximately equal distribution of women and men are extremely rare; the small section *Mathematics Education and Popularization of Mathematics* at the last ICM in Rio de Janeiro with two talks by women and one talk by a man seems to be the only example so far. Nevertheless, the presence of at most a single woman among the section speakers as criticized by Sadosky remains by far the usual practice.

6 Discussion

In our study we have used the complete list of plenary and section speakers invited to ICMs from its beginning in 1897 through the most recent congress in 2018 in Rio de Janeiro. We have combined this list with demographic information from Wikidata and section names of the talks. We have further enriched our data, in particular by inferring the gender. Our data is provided at [12] and can be used for further research.

The participation of women in the International Congresses of Mathematicians has increased over the course of the past 121 years. However, as shown in Fig. 3, their share does not show a clear trend over time. Instead, the inclusion of women among ICM speakers shows a clear peak in the late 1920s and early 1930s of the last century, followed by a long period of almost complete absence, until the start of a continuous positive development in the late 1980s that persists until today. As noted by Izabella Laba with respect to the equitable representation of women speakers at ICM-2014, "Compared to the proportion of women among tenured and tenure-track faculty at research universities, the group from which ICM invited speakers usually hail and therefore a more appropriate benchmark, it does not look far off" [10].

We have analyzed the relation of the speakers' gender with their countries of citizenships and the topics of the sections in which they presented their talks. As expected, most speakers came from a rather small set of countries, while many countries and even entire continents were barely represented at all. The distribution of citizenships with respect to gender, however, does not show significant difference. More than 70% of all women and men speakers are or were citizens of countries whose territory today corresponds to the United States, France, Russia, Germany, United Kingdom and Italy. We have further shown that the selection of speakers reflects a certain regional focus, yielding, in most cases, a noticeably higher share of speakers originating from the congress's host country. This effect is particularly pronounced for countries that are overall well represented among ICM speakers. For women, however, this tendency is not so clear and would require further investigation. Cost and ease of travel as well as the language are likely to have played a role in the decisions to invite a speaker, in particular in the early congresses. Nevertheless, it would be interesting to explore the possible relations between the nationalities of the members of the Programme Committee and of the sectional committees with the nationalities of the invited women (and men).

The individual congresses show high variance in the arrangements of talks into sections, making a grouping of sections inevitable. We have arranged the non-plenary talks since ICM-1970 into eight large groups and studied the distribution of women across these. We have found that the share of women ranges between less than 5% in sections related to Algebra and Number Theory to more than 16%

in sections on Logic, Foundations, Philosophy, History and Education. Talks by women sum up to 6% of all talks in the section group Analysis, ODEs, PDEs, Dynamical Systems which is by far the largest one. These effects are not consistent with the representation of women as authors in the respective fields. However, a correlation on this level might not be very likely anyway since ICM invitations are very rare and, in some sense, singular (for women). An analysis of gender and topics among authors in highly prestigious journals or among tenured and tenure-track faculty at research universities would hence presumably constitute a more suitable set in order to understand whether the fluctuations between ICM sections might partially be explained by a lack of 'suitable' women in the respective fields. It should, however, be noted that the breakdown by sections leads to a rather small number of individuals per group, which is more prone to variation.

It would be interesting to consider the longitudinal development of women's participation in other conferences in mathematics (and other fields) as well. Smaller conferences might constitute a much bigger issue in terms of inclusion since, as claimed in [10], there are no large committees overseeing them. In recent years, (women) scientists in STEM fields have proposed the formulation of a Bechdel Test, a measure of women's representation in fiction (movies, comics, video games etc.), for scientific workshops. A movie would pass the Bechdel test if (1) it features at least two named women, (2) who talk to each other, (3) about something besides a man. An analogous test for scientific conferences could require (1) at least two female invited speakers, (2) who are asked questions by female audience members, (3) about their research.[10] As noted by various scientists in social media channels, this form of test is rarely passed by conferences in STEM fields; even among the recent ICMs, many sections would not pass this test either. Such a test, while measuring only a basic level of inclusion of women and despite being far from creating a 'critical mass' in the respective conferences, would yield a first understanding of women's participation in mathematical conferences across fields and over time.

Manifold factors have played a role in the longitudinal evolution of the number of ICM lecturers at ICM. Of particular importance for the sustained positive evolution in the last decades was and is the establishment of various associations of women in mathematics and their efforts to increase the visibility of their female colleagues in the field. As suggested by Elena Resmerita and Carola-Bibiane Schönlieb, the current convenors of the European Women in Mathematics (EWM), it is crucial to keep highlighting contributions of women mathematicians by continuing to showcase contributions of women in mathematics, "raise the profile of women mathematicians, volunteer to serve on committees of international mathematical associations and mathematical award committees, nominate female colleagues or

---

[10]http://openscience.org/a-bechdel-test-for-scientific-workshops/.

encourage others to nominate them, and overall help to build a scientific atmosphere without boundaries."[11]

Acknowledgements This research was conducted as part of the project 'A Global Approach to the Gender Gap in Mathematical, Computing, and Natural Sciences: How to Measure It, How to Reduce It?', supported by the International Science Council and several scientific unions. We thank June Barrow-Green, Fulvia Furinghetti, Catherine Goldstein and Annette Vogt for providing important insight and expertise from History of Mathematics. We further thank Lucia Santamaria for suggestions that greatly improved the manuscript. We gratefully acknowledge the support of the International Mathematical Union for granting us access to their data on ICM Plenary and Invited Speakers.

Appendix

Table 1 Percentage of all speakers and all women speakers, respectively, whose country of citizenship equals the host country of the respective congress

Year	Host country	Speakers from host country (%)	Women speakers from host country (%)
1897	Switzerland	3.1	0
1900	France	32.4	0
1904	Germany	45.2	0
1908	Italy	32.7	100
1912	United Kingdom	26.5	100
1920	France	31.4	0
1924	Canada	2.3	0
1928	Italy	25	50
1932	Switzerland	7.4	8.3
1936	Norway	6.1	0
1950	United States of America	36.6	0
1954	Netherlands	2.5	0
1958	United Kingdom	6.9	0
1962	Sweden	2.9	0
1966	Russia	30.4	100
1970	France	8.1	0
1974	Canada	2.3	0
1978	Finland	0.5	0
1983	Poland	3	0

(continued)

---

[11] Statement by the EWM convenors Elena Resmerita and Carola-Bibiane Schönlieb in reaction to the absence of women among the Fields medalists in 2018. https://www.europeanwomeninmaths.org/ewm-statement-about-icm-2018.

Table 1 (continued)

Year	Host country	Speakers from host country (%)	Women speakers from host country (%)
1986	United States of America	39.0	40
1990	Japan	10.1	0
1994	Switzerland	1.7	0
1998	Germany	7.6	0
2002	China	3.3	5.3
2006	Spain	3.8	0
2010	India	7.7	3.6
2014	South Korea	0	0
2018	Brazil	5.9	12.5

References

1. Andler, M.: Who are the invited speakers at ICM 2014? Eur. Math. Soc. Newsl. 92, 38–44 (2014)
2. Barrow-Green, J., Gray, J.: Geometry at Cambridge, 1863–1940. Historia Mathematica 33, No. 3, 315–356 (2006) doi: 10.1016/j.hm.2005.09.002
3. Case, B.A., Leggett, A.M.: Across borders. In: Case, B.A., Leggett, A.M. (eds.) Complexities: Women in mathematics, pp. 121–128. Princeton, NJ: Princeton University Press (2005)
4. Fields, J.C. (ed.): Proceedings of the International Mathematical Congress (ICM), August 11–16, 1924, Toronto, Canada. Toronto: The University of Toronto Press (1928). https://www.mathunion.org/fileadmin/ICM/Proceedings/ICM1924.2/ICM1924.2.ocr.pdf
5. Furinghetti, F.: The emergence of women on the international stage of mathematics education. ZDM Mathematics Education 40:529–543 (2008). doi: 10.1007/s11858-008-0131-y
6. Gutzmer, A. (ed), Jahresber. Dtsch. Math.-Ver. 15, pp. 12 (1906). https://gdz.sub.uni-goettingen.de/id/PPN37721857X_0015
7. ICM Plenary and Invited Speakers | International Mathematical Union (IMU). Mathunion.org. available at https://www.mathunion.org/icm-plenary-and-invited-speakers [accessed January 3, 2018]
8. Israel, G.: Italian Mathematics, Fascism and Racial Policy. In: Emmer, M. (ed.) Mathematics and Culture I, pp. 21–48. Heidelberg: Springer (2004)
9. IPSIA Verona. Giovanni Giorgi. available at http://www.giorgivr.it/pvw/app/VRIP0003/pvw_sito.php?sede_codice=VRIP0003&page=1940003. [accessed October 13, 2018]
10. Laba,I.: Gender, conferences, conversations and confrontations. The accidental mathematician 15.3.2015. https://ilaba.wordpress.com/2015/03/15/gender-conferences-conversations-and-confrontations/ [accessed September 9, 2018]
11. List of International Congresses of Mathematicians Plenary and Invited Speakers. En.wikipedia.org. available at https://en.wikipedia.org/w/index.php?title=List_of_ International_Congresses_of_Mathematicians_Plenary_and_Invited_Speakers [accessed February 12 2018]
12. Mihaljevič, H.: Plenary and Invited Speakers of the International Congress of Mathematicians (ICM) [Data set]. Zenodo. (2018) http://doi.org/10.5281/zenodo.1976747

13. Mihaljevič-Brandt, H., Santamarìa, L., Tullney M.: The Effect of Gender in the Publication Patterns in Mathematics. PLOS ONE 11(10): e0165367 (2016) doi: 10.1371/journal.pone.0165367
14. Moore, C.L.E.: The fourth International Congress of Mathematicians: sectional meetings. Bull. Amer. Math. Soc. 15, 8–43 (1908). http://www.ams.org/journals/bull/1908-15-01/S0002-9904-1908-01685-9/S0002-9904-1908-01685-9.pdf.
15. Pisati Laura – Scienza a due voci. Scienzaa2voci.unibo.it. available at http://scienzaa2voci.unibo.it/biografie/1166-pisati-laura. [accessed September 11, 2018]
16. Sadosky, C.:Affirmative action: what is it and what should it be?. In: Case, B.A., Leggett, A.M. (eds.) Complexities: Women in mathematics, pp. 116–120. Princeton, NJ: Princeton University Press (2005)
17. Santamarìa, L., Mihaljevič, H.: Comparison and benchmark of name-to-gender inference services. PeerJ Computer Science 4:e156 (2018) doi: 10.7717/peerj-cs.156
18. Satake, I. (ed.): Proceedings of the International Congress of Mathematicians (ICM), August 21–29, 1990, Kyoto, Japan. Volume I. Tokyo etc.: Springer-Verlag (1991). https://www.mathunion.org/fileadmin/ICM/Proceedings/ICM1990.1/ICM1990.1.ocr.pdf
19. zbMATH - the first resource for mathematics. Zbmath.org. Author profile of L. Pisati. available at https://zbmath.org/authors/?q=ai%3Apisati.l [accessed July 20 2018]

The Historical Context of the Gender Gap in Mathematics

June Barrow-Green

Abstract This chapter is based on the talk that I gave in August 2018 at the ICM in Rio de Janeiro at the panel on *The Gender Gap in Mathematical and Natural Sciences from a Historical Perspective*. It provides some examples of the challenges and prejudices faced by women mathematicians during last two hundred and fifty years. I make no claim for completeness but hope that the examples will help to shed light on some of the problems many women mathematicians still face today.

1 Introduction

In 1971 the *Association for Women in Mathematics* (AWM), the first organisation for supporting women in mathematics, was established in the United States.[1] There are now many organisations worldwide supporting women in mathematics, and the number continues to grow, with the IMU's Committee for Women in Mathematics (CWM) providing a focal point.[2] Nevertheless, despite the extensive work that has been done since 1971 to address the particular challenges which confront women in mathematics, women still face particular difficulties within their professional careers. Many of these difficulties have a long history stemming from deeply embedded cultural attitudes, some of which have proven difficult to shift. The fact that these attitudes have a long history does not excuse why change has been slow but it does perhaps help to explain it.

In what follows I describe some examples of the challenges faced by women mathematicians during the last two hundred and fifty years, looking first at some individuals from the eighteenth and nineteenth centuries, then taking a slightly

---

[1] For a history of the AWM, see [1].

[2] See https://www.mathunion.org/cwm/organizations/country.

J. Barrow-Green (✉)
The Open University, Milton Keynes, UK
e-mail: june.barrow-green@open.ac.uk

© The Association for Women in Mathematics and the Author(s) 2019
C. Araujo et al. (eds.), *World Women in Mathematics 2018*, Association for Women in Mathematics Series 20, https://doi.org/10.1007/978-3-030-21170-7_6

broader view and considering women within particular national contexts.[3] I make no claim for completeness but through these historical examples I hope to shed light on some of the problems still encountered today.

2 The Eighteenth and Nineteenth Centuries

The first woman in the modern period to make a substantial contribution to mathematics was the Italian Maria Agnesi (1718–1799) who in 1748 published one of the earliest textbooks on the differential and integral calculus, *Instituzioni Analitiche*, which she originally wrote in order to instruct her younger brothers. Written in the vernacular (which was unusual in the period), the book was accessible to a broad audience and an important contribution to the spread of the calculus in Italy. Two years after the book's publication, she was appointed to the chair of mathematics in Bologna on the recommendation of the Pope, Benedict XIV, but she never took up the position. Agnesi did not even go to Bologna although her name remained on the rolls of the university. Instead she devoted her life to works of charity.[4]

Much has been written on the content and reception of Agnesi's text but I want to draw attention only to some remarks made by the French historian of mathematics, Jean-Etienne Montucla, as they are illustrative of contemporary views about women mathematicians. Montucla, who was writing at the end of the eighteenth century, was complimentary about the book but nevertheless rued the fact that there was no translation of it by one of the French women mathematicians—he didn't say who he had in mind—thereby implying that he believed there to be a difference between the way men and women approach and study mathematics.[5] At the same time, he was also astonished that a woman—or as he put it "a person of a sex that seems so unfit to tread the thorny paths of abstract sciences" [3]—could penetrate so deeply into the calculus, thereby reinforcing the notion of the general unsuitability of women for mathematics.

Agnesi, along with a number of other women in the eighteenth and early nineteenth century, such as Émilie du Châtelet (1706–1749), Ada Lovelace (1815–1852) and Mary Somerville (1780–1872), all of whom made lasting and significant

---

[3]By women mathematicians I mean women who were producing or developing original mathematics. That is not to diminish the contribution of the many women who simply used mathematics, as for example in accounting practices, or who were employed as human computers, but simply to note that they are not the subject of my discussion.

[4]For a discussion of Agnesi's life, see Paula Findlen's excellent essay review [2].

[5]There was a French translation by a man, Pierre Thomas d'Antelmy, which appeared in 1775. An English translation by John Colson appeared in 1801. It was a mistranslation by Colson, who confused *a versiera* (the rope that turns a sail) with *l'aversiera* (she-devil), that led to the cubic curve studied by Agnesi being named the 'witch of Agnesi' (an early example perhaps of unconscious bias?).

Fig. 1 "Mlle Ferrand
méditant sur Newton" by
Maurice-Quentin de La Tour

contributions to mathematics, were not prevented from doing mathematics, in fact sometimes rather the opposite. For example, Lovelace, today renowned for her remarkable paper which explained the principles of Charles Babbage's analytical engine [4], was encouraged by her mother to study mathematics with Augustus De Morgan.[6] Something these women all had in common was that they came from a social class which gave them the time and the opportunity to discuss mathematics (and natural philosophy) with men on equal terms. Both Somerville and Lovelace attended Babbage's scientific soirées and together they frequently called on him in order to see and to discuss his analytical engine.

That Élisabeth Ferrand (1700–1752), an important influence on Abbé de Condillac and a friend of Alexis Claude Clairaut, chose a page from Voltaire's influential *Éléments de la philosophie de Newton* (1738)—the book which introduced Newtonian physics to France—as the backdrop to her portrait is indicative of such learning among women in Enlightenment circles (Fig. 1).[7] But there may be another reason Ferrand chose Voltaire's book; for Voltaire was not its sole author although his is the only name to appear on the cover and title page. Voltaire's long-time companion, Émilie du Châtelet, played a major role in the book's production and in fact Voltaire did not shy from acknowledging it. Du Châtelet's name appears twice in the introductory matter where Voltaire gives an indication of their collaboration, and

---

[6]The teaching was done mostly by correspondence, see [5].

[7]For biographical information about Ferrand and a discussion of Maurice-Quentin de La Tour's pastel portrait, see http://www.pastellists.com/Essays/LaTour_Ferrand.pdf.

Fig. 2 The frontispiece to
Voltaire's *Éléments de la
Philosophie de Newton*

she is also depicted in the frontispiece where he has imagined her as a muse floating
above him while holding a mirror reflecting Newton's wisdom down onto his hand,
thus implicitly admitting her scientific superiority (Fig. 2). Although co-authoring
was not common practice at the time, by keeping du Châtelet's name from the front
cover, Voltaire was nevertheless diminishing the visibility of women as genuine
contributors to serious scientific work. Some ten years later du Châtelet completed
her own much more ambitious work: *Principes Mathématiques de la Philosophie
Naturelle*, a translation from the Latin of Isaac Newton's fiercely difficult *Principia*.
But it was much more than a translation: Newton's geometry was rendered into
algebra and she provided an extensive commentary including recent research. She
completed it while pregnant and died shortly after giving birth. It was not published
until 1759, ten years after her death, the publication timed to coincide with the year
of the return of Halley's Comet. Today it is still the only complete translation into
French of the *Principia*, a testimony to du Châtelet's ability as a mathematician.

Although it was acceptable for these women to mix socially in mathematical and scientific circles, they could not hold any sort of academic or institutional position. Somerville was able to make money from the sales of her books—her *Mechanism of the Heavens* (1831), an acclaimed translation and commentary on the celestial mechanics of Pierre-Simon Laplace, became a recommended text for men studying for the Mathematical Tripos at Cambridge[8]—and she could have a paper published by the Royal Society of London,[9] but there was no question of her being admitted as a Fellow of the Society. She could not present her paper to the Society: that had to be done by her husband, William, who was a Fellow. Although that is not to say that the Society did not recognise her scientific excellence. In 1842 HRH The Duke of Sussex (then the most recent past President of the Society) together with other subscribers presented the Society with a marble bust of Somerville to be placed in the Great Hall.

More than a century would elapse before women would be admitted as Fellows of the Royal Society. In 1902 when the physicist Hertha Ayrton (1854–1923) was formally proposed as a candidate for Fellowship for her pioneering work on the electric arc, one reason for not admitting her was the fact that she was married, and married women had no status in law! Although the Royal Society would not admit Ayrton as a Fellow, in 1904 they did allow her to read a paper before the Society— the first woman to do so—and in 1906 they awarded her the Hughes Medal.[10] Thus, the Fellows of the Society were prepared to acknowledge that women could do science, and indeed do it very well, but they were not prepared to accept that women should or could be considered as their scientific equals. It would be another forty years before they would change their minds. The first women to be admitted to the Royal Society were admitted in 1945; the first woman mathematician, Mary Cartwright (1900–1998), was admitted in 1947 [8].

The first woman to be a professional academic mathematician in the modern sense was the Russian Sofia Kovalevskaya (1850–1891). Championed by the Swedish mathematician Gösta Mittag-Leffler, who overcame strong opposition to secure her appointment at the Stockholm Högskola (the forerunner of Stockholm University), she became a full professor in 1889. But despite Kovalevskaya's

---

[8]On 14 February 1832, George Peacock, who in 1837 would become the Lowndean Professor of Geometry and Astronomy at Cambridge, wrote to Somerville to say that he considered *The Mechanism of the Heavens* "to be a work of the greatest value and importance," and told her that "Dr Whewell and I have already taken steps to introduce it into our course of studies at Cambridge and I have little doubt that it will immediately become an essential work to those of our students who aspire to the highest places in our examinations." [6].

[9]The paper [7] was Somerville's first publication and although the conclusions in it were later disproved, it established her as a practitioner of science rather than as a student or an onlooker.

[10]Ayrton was the fifth recipient of the Hughes Medal—awarded "in recognition of an original discovery in the physical sciences, particularly electricity or magnetism or their applications"— and the first woman to be awarded it. To date it has been awarded to only one other woman: Michele Dougherty in 2008.

internationally recognised mathematical talent—she was awarded the Prix Bordin[11] of the Académie des Sciences in Paris in 1888 for her work on the spinning top, with the prize money being increased from 3000 to 5000 francs due to the originality of her results—there was no chance for her to gain a position in one of the mathematical centres of Europe, such as Paris or Berlin [9]. As a foreigner it would have been struggle enough but being a woman made it impossible.

Kovalevskaya herself reported examples of the prejudice that she encountered. In 1869, early in her career, when she was making one of her visits to the London salon of the novelist George Eliot (Mary Anne Evans) she found Eliot, who had an interest in mathematics,[12] very keen to introduce her to the philosopher Herbert Spencer because, as Eliot said openly on the occasion, Spencer denied "the very existence of a woman mathematician" [10]. Then, later, in December 1884, shortly after her appointment as an assistant professor in Stockholm, she would write to Mittag-Leffler [11]:

> I have received from your sister, as a Christmas present, an article by Strindberg, in which he proves as decidedly as two and two make four, what a monstrosity is a woman who is a professor of mathematics, and how unnecessary, injurious, and out of place she is.

These episodes provide a stark reminder of the fact that this was a period when it was widely believed that if women used their brains in order to do mathematics (or science), the effort involved would put a strain on their physical well-being, sapping their strength to such an extent that it would interfere with their ability to have children.

As a gifted female mathematician, Kovalevskaya inevitably attracted attention, but not only because of her mathematics. In 1886, Charles Hammond, the assistant of the English mathematician, James Joseph Sylvester, on seeing a photograph of Kovalevskaya, declared that she was "the first handsome mathematical lady" he had ever seen [12]. (Of course one can wonder how many mathematical ladies he had ever seen!) Clearly beauty was not expected in a female mathematician. After Kovalevskaya's untimely death—she died unexpectedly aged only 41—interest in her appearance intensified. But no longer was there a consensus—for some she was beautiful for others she was not and there was no general agreement. The differing nature of these opinions provides an insight into the disparate views about female mathematicians [13].

Although Kovalevskaya's standing as a mathematician was high at the time of her death, it later suffered setbacks. It is true that some errors were found in her work but nothing that would have harmed her reputation had she been a man. One of the most egregious examples came from the pen of the Italian mathematician

---

[11]The Prix Bordin, which was second in prestige to the Grands Prix of the Académie, was awarded for scientific subjects as well as mathematics.

[12]Eliot attended geometry lectures in London and frequently incorporated mathematics into her novels, notably *The Mill on the Floss*.

Gino Loria, professor of mathematics in Genoa, who in 1903 was putting forward the case for keeping mathematical faculties closed to women [14]:

> As for Sophie Germain and Sonja Kowalevksy, the collaboration they obtained from first-rate mathematicians prevents us from fixing with precision their mathematical role. Nevertheless, what we know allows us to put the finishing touches on a character portrait of any woman mathematician: She is always a child prodigy, who, because of her unusual aptitudes, is admired, encouraged, and strongly aided by her friends and teachers; in childhood she manages to surpass her male fellow-students; in her youth she succeeds only in equalling them; while at the end of her studies, when her comrades of the other sex are progressing, fresh and courageous, she always seeks the support of a teacher, friend or relative; and after a few years, exhausted by the efforts beyond her strength, she finally abandons a work which is bringing her no joy …

As Roger Cooke has observed, the most malign element of Loria's judgement is his implication of the necessity of "fixing with precision" the originality in a woman's work before admitting it is good [15]. The clear insinuation here being that Weierstrass, Kovalevskaya's teacher, had more to do with Kovalevskaya's work than was apparent, despite the fact that Kovalevskaya was meticulous about citing him. Happily, more recent scholarly work has restored Kovalevskaya to her rightful place in the mathematical pantheon.[13]

Loria's reference to Sophie Germain (1776–1831) prompts some further remarks. Germain, who initially taught herself mathematics from books in her father's library, at the age of eighteen began to read the lesson-books of the professors at the École Polytechnique. As the École did not admit women and she wanted to take her mathematics further, she struck up a correspondence with one of the professors, Joseph-Louis Lagrange. But she used the name of a real male student, fearing, as she later said, "the ridicule attached to a female scientist." She subsequently used the same pseudonym when corresponding with Gauss. However, on discovering her true identity, neither Lagrange nor Gauss responded adversely. Indeed in both cases they were complimentary. She had impressed them with her mathematics and to them that was what mattered. In 1807 Gauss wrote to Germain [17]:

> But when a woman, because of her sex, our customs and prejudices, encounters infinitely more obstacles than men in familiarizing herself with their knotty problems, yet overcomes these fetters and penetrates that which is most hidden, she doubtless has the most noble courage, extraordinary talent, and superior genius.

Notwithstanding the fact that both Lagrange and Gauss were impressed by Germain, and that her work on elasticity gained a prize from the Paris Académie des Sciences,[14] after her death, like Kovalevskya, Germain's star lost much of its lustre. The view that women were not suited to, and therefore not capable of, doing work in higher mathematics, dominated. In Germain's case, this is particularly apparent in connection with her work on Fermat's Last Theorem where her contribution steadily became conflated with that of Adrien-Marie Legendre who credited her with only a small part of a much larger and more substantial piece of work. It was not until the

---

[13] See, for example [9, 16].

[14] Germain was the first woman to gain such a prize.

1990s when David Pengelly and Reinhard Laubenbacher worked on her manuscripts and letters that the true scale of her contribution to the problem was recognised. At the end of their research, Pengelly and Laubenbacher concluded [18]:

> Sophie Germain was a much more impressive number theorist than anyone has ever known.

3   Cambridge University

During the nineteenth century, Cambridge was the beating heart of British mathematics and the Mathematical Tripos the most prestigious and demanding examination in Britain. It was punishingly hard, both physically and mentally, but the rewards were great. Students who came high up in the list of wranglers (students in the first class) had a passport to the career of their choice, be it the law, medicine, the church or mathematics. It is hard to over-estimate the kudos attached to being senior wrangler, the top mathematics student of the year. Kudos that went far beyond the bounds of Cambridge.

From the second half of the century, women could study mathematics at Cambridge—the women's colleges Girton and Newnham were founded in 1869 and 1872 respectively—but they had to obtain permission to sit the Tripos examination, they could not do so by right, and they could not be awarded a degree (with its privileges and voting rights). For over three-quarters of a century the two colleges were not even officially part of the University (that had to wait until 1948).

In 1880 Charlotte Scott (1858–1931) created a sensation by being judged equal to the 8th wrangler.[15] The newspapers and periodicals were full of her success—she had done better than 93 of the 102 men taking the examination—and the reports are revealing about contemporary views of women mathematicians. *The Spectator* is typical [19]:

> Miss Scott has answered papers set for the mathematical tripos in a manner which would have brought her high on the list of Wranglers, an achievement of no common kind. ... We hope that the ability which the new system brings out and fosters in women, will not be of a kind to give to those who possess it a character for deficiency in feminine gentleness. We do not believe that it will be so. But even in the rare cases where it is so, the world should remember that there have always been women of the masculine type—only that they have hitherto lacked the means of proving what they could do, though possessing amply the means of proving what they could not be.

Once again mathematics is portrayed as an essentially male pursuit. Nevertheless, Scott's achievement generated a growth in support for female students, and from 1881 women were given the right to take the examinations and to have their results published, albeit separate from the men. But they still could not be awarded degrees.

An even greater sensation was created when, in 1890, Philippa Fawcett (1868–1948) was judged to be above the senior wrangler. Reports were published far

---

[15]1880 was a strong year with Joseph Larmor, future Lucasian professor, being senior wrangler, and J. J. Thomson, future Nobel laureate, being second wrangler.

Fig. 3 Philippa Fawcett
celebrated in *Punch*, 21 June
1890

"SENIORA FAWCETT."
So to be entitled henceforth, as she is Seniorer to the Senior Wrangler.

and wide, including in the *New York Times* [20]. The satirical magazine *Punch* even produced a cartoon (Fig. 3). Fawcett had scored 13% more marks than the highest ranked man, G.T. Bennett, and achieved what many had believed impossible. Nevertheless, when the Tripos list was published, her name (together with that of the other women) still appeared below that of all the men, her position in the examination "above the Senior Wrangler" written in brackets beside it.

After Fawcett's success, the clamour for women to be awarded degrees grew louder but it was still not loud enough. Cambridge did not fully open its doors to women until December 1947. Those who wanted degrees had to go to London or, from 1920, Oxford.

Grace Chisholm (1868–1944), who was placed between the 23rd and 24th wranglers in the Mathematical Tripos of 1892 (and also unofficially achieved a first class in the Final Honours School of Mathematics in Oxford the same year), completed her studies with Felix Klein in Göttingen (see section on Germany below).[16] As a

---

[16]For a description of Chisholm's experience in Göttingen, see [21].

woman, there was nowhere in Britain she could engage in post-graduate research.[17] In 1895 she became not only the first British women to gain a PhD in mathematics but also the first woman anywhere to gain one following a standard period of study and oral examination.[18] Shortly afterwards she married the mathematician William Henry Young who had been her tutor for a term at Girton [22]. Young was content for her to continue with mathematical research but, as he told her, publishing mathematical papers was a man's game [23]:

> The fact is that our papers ought to be published under our joint names, but if this were done neither of us get the benefit of it. No. Mine the laurels now and the knowledge. Yours the knowledge only. ...Everything under my name now, and later when the loaves and fishes are no more procurable in that way, everything or much under your name.

In the end, the Youngs published 214 papers between them [24]. They were mostly published in William's name with only 18 in Grace's name and 13 co-authored. In 1906 they published their book *The Theory of Sets of Points* together. Although happily such a situation no longer pertains, that is not to say that publishing for women today is without problems. Recent analysis of mathematical publications dating from 1970 has shown that "gender-related publication patterns exist and are one of the factors that lead to an underrepresentation of women in mathematics" [25].

Returning to the situation at Cambridge, aside from Young there were men there who were prepared to provide at least some support for women mathematicians. Charlotte Scott studied algebraic geometry with Arthur Cayley, the Sadleirian Professor, and it was Cayley who recommended her for the position of head of mathematics at the newly opening Bryn Mawr College in the United States, a position which she took up in 1895, no equivalent opportunity being available to her in Britain. Cayley was an active supporter of women's education—for several years he was president of the Council of Newnham College—but realizing that change would be achieved only slowly his work was chiefly behind the scenes. For a long time men like Young and Cayley were part of a small minority in Cambridge—the belief that women were not capable of doing serious mathematics proved extremely hard to shift.

After 1947, women may have been able to get degrees at Cambridge but little changed in other respects, and progress towards gender equality in mathematics has been glacially slow. Mary Cartwright, despite being a Fellow of the Royal Society, was never deemed worthy of a professorship.

The first woman to be elected to a professorship in Cambridge was the applied mathematician Anne Davis in 2002, and as at 2018 a female professor in pure

---

[17]Men in Britain could engage in post-graduate research but if they wanted a PhD they had to go abroad. The PhD did not come to Britain until after the First World War.

[18]Kovalevskaya, who had studied privately with Weierstrass in Berlin, was awarded a PhD in absentia from Göttingen in 1874 based on the contents of three separate research papers. She took no oral examination.

mathematics has yet to be appointed there. Furthermore, there is still a greater gender imbalance among mathematics students at Cambridge than at other universities.[19]

4 Germany

In 1764 Immanuel Kant had pronounced that women who succeeded in mathematics "might as well have a beard" [26]. His point being that if women did succeed in mathematics then they would not be women they would be men! The first concrete sign of progress was in 1874 when Kovalevskaya, having studied privately with Karl Weierstrass in Berlin, was awarded a PhD by the University of Göttingen. But it remained an isolated incident until the 1890s when Felix Klein, and subsequently David Hilbert, in Göttingen were allowed to let women to audit lectures. Initially only foreign women were permitted as their special status meant they could be used as a test case.[20] As Klein observed in 1895, "The opinion still prevailing in Germany is that the study of mathematics must be as good as inaccessible to women" [27]. At that time he himself had had six foreign women—American, English (including Grace Chisholm) and Russian—successfully participating in his higher mathematics lectures which prompted him to remark [27]:

> No one would wish to assert, however, that these foreign nations possess some inherent and specific talent that evades us, and thus that, with suitable preparation, our German women should not be able to accomplish the same thing.

Klein encouraged his women students to publish in *Mathematische Annalen*, the journal of which he was the chief editor. The American Mary Winston (1869–1859), whom Klein had originally met in 1893 when she attended both the Mathematics Congress in Chicago and his Evanston Lectures that followed it, was the first, in 1895, with a short note on the hypergeometric function.

The most prolific female author in *Mathematische Annalen* under Klein's editorship was unsurprisingly Emmy Noether (1882–1935), one of the twentieth century's most gifted mathematicians. Noether's life and extraordinary talent for mathematics have been well documented but recently more information has come to light with regard to her unsuccessful application in 1928 for a professorship at Kiel, information which underlines the tremendous difficulty and prejudice she faced in trying to get a position in Germany. When Noether's name was suggested

---

[19]For a discussion about the current situation with respect to mathematics students in Cambridge, see the Varsity interview of 2 November 2017 with Julia Gog https://www.varsity.co.uk/news/ 13945. In 2014 the Faculty of Mathematics at Cambridge achieved an Athena SWAN bronze award. See https://www.maths.cam.ac.uk/womeninmaths/athenaswan.html.

[20]In 1891 the American Ruth Gentry was permitted to audit the lectures of Lazarus Fuchs and Ludwig Schlesinger in Berlin for one term before permission was revoked.

as a possible contender for the professorship by Adolf Fraenkel, a professor at Kiel, Helmut Hasse, then a professor in Halle, was moved to say [28]:

> I am astonished you even seriously consider this possibility. Although I regard her highly in scientific matters, I deem her totally unfit to fill a regular teaching position, even less so in a small university like Kiel. ... I am of the opinion that one should not make the experiment to appoint a woman as full professor at such a place as Kiel. The experiment should be tried first on a bigger scale where an unsuccessful outcome would not do so much harm.

The applied mathematician Theodor Kaluza was appointed to the position. In 1933 Noether, as a Jew, was dismissed from her "extraordinary" professorship in Göttingen (basically a Privatdozent with an additional small stipend), and emigrated to the United States where she had a temporary position at Bryn Mawr until her premature death in 1935.

5   United States of America

Thanks to the detailed work of Judy Green and Jeanne LaDuke there is now a wealth of information available about the 228 American women mathematicians who earned PhDs in the United States before 1940 [29].[21] Added to that is the research by Sarah Greenwald, Anne Leggett and Jill Thomley on the AWM which brings the picture in the United States almost up to date [30]. What is striking about the latters' findings is how the percentage of women mathematics PhDs rose fairly steadily decade on decade from the end of the nineteenth century up to the beginning of the Second World War only then to drop off significantly. As can be seen from the graph which Greenwald et al produced (Fig. 4), the 1930s percentage was only really surpassed in the 1990s.

In the pre-WW2 period, certain institutions in the United States stand out with respect to their support for women mathematicians. Bryn Mawr, the women's college founded in 1895, benefited from having Charlotte Scott at its mathematical helm. Scott supervised seven PhD students and her colleague, and successor as head of mathematics, Anna Johnson Pell Wheeler (1883–1966) supervised six. Both of them, together with Olive Hazlett (1890–1974) who spent a short time as a lecturer at Bryn Mawr, are distinguished for being the only starred women mathematicians in (the inaccurately named) American Men of Science between 1903 and 1943. At the University of Chicago, Leonard E. Dickson supervised 18 women PhDs (27% of his output), and Gilbert A. Bliss supervised 12 women PhDs (23% of his output). Meanwhile at Cornell, Virgil Snyder supervised 14 women PhDs (37%) of his output. In addition, as mentioned above, Klein in Göttingen also supported American women mathematicians.

---

[21] For additional material by the same authors, see http://www.ams.org/publications/authors/books/postpub/hmath-34.

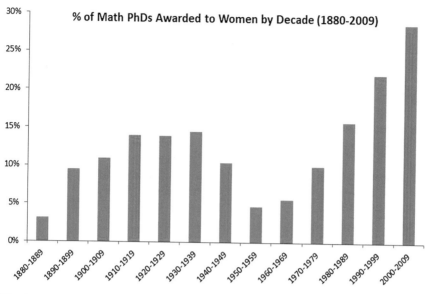

Fig. 4 Percentage of Math PhDs Awarded to Women by Decade (1880–2009)

The lowering of numbers in the immediate post-WW2 period can be largely attributed to the prevailing social conditions which conspired against women mathematicians as it did against women in other fields. It was not until the 1970s, with the advent of organisations supporting women mathematicians, that significant improvements were made.

6 The Growth of Institutional Support for Women in Mathematics

In general, national mathematical societies have been welcoming to women members. However, the same cannot be said of their governing bodies. The American Mathematical Society was exceptional in appointing Charlotte Scott as a Vice-President in 1906, but it took the Society until 1983 before it appointed its first woman president, Julia Robinson. The first Society to appoint a woman president was the Société Mathématique de France when they elected Marie-Louise Dubreil-Jacotin in 1952. Even in recent times, the number of women in senior roles within societies has not accurately reflected the contribution of women to mathematics as a whole.

After the formation of the AWM in 1971, a number of other organisations supporting women in mathematics were established in North America and Europe: The Joint Committee on Women in the Mathematical Sciences (1971), European Women in Mathematics (1986), The Women in Mathematics Committee of the

European Mathematical Society (1991), Femmes et Mathématiques (1987), The Canadian Society Committee for Women in Mathematics (1992) and the London Mathematical Society Women in Mathematics Committee (1999).

At the First European Congress of Mathematics in 1992, there was a Round Table on Women in Mathematics organised by the Women in Mathematics committee (WiM) of the European Mathematical Society. The aim of the Round Table was to look at the proportion of women involved in mathematics in various countries. Its report contains a wealth of information and data providing a detailed picture of the situation [31]. Among the examples demonstrating the ingrained bias that still existed at that time was one provided by Eva Bayer-Fluckiger concerning the pre-printed postcards supplied by mathematical departments to be used for reprint requests. These cards contained a text that ran roughly as follows:

Dear Sir,
Please send me a reprint of your article . . . that appeared in. . . .
Many thanks in advance.
Yours sincerely.

Bayer-Fluckiger had recently received such a card that had the text in three languages: "Lieber Herr Professor; Monsieur le Professor; Dear Sir"Ï. The idea that the author of a mathematical article could be a woman simply wasn't entertained. In commenting on the "deplorable" German situation in comparison with other European countries, Christine Bessenrodt, in her talk at the Round Table, noted that although about one third of students in mathematics were women, "only 9% of dissertations and 7% of habilitations are written by women, and only 2% of the professors in mathematics are female." And she found it a very depressing task "investigating the status quo and compiling a list of obstacles that women in mathematics have to face in Germany." So much so that she found "it surprising that there are women mathematicians at all in [Germany]"Ï. Such examples paint a vivid picture of the poor situation for women mathematicians in Europe at that time as well as highlighting the extent of the gender gap.

The WiM committee, helped by funding from the UK Royal Society Athena Awards, gathered data about women mathematicians across Europe in 1993 and again in 2005 (Fig. 5).[22] Although the data shows a substantial increase in the percentage of women mathematicians during the intervening 12 years, it also reveals a significant difference between north and south, highlighting the countries in which the greatest efforts need to be made.

Since 2000 the number of organisations set up to support women in mathematics has grown worldwide. In addition to those organisations named above, there are now organisations in countries across the globe, as well as umbrella organisations for African Women in Mathematics and Central Asian Women in Mathematics.[23]

---

[22]For details of the data, see C. Hobbs, E. Koomen 'Statistics on Women in Mathematics' (2006), https://womenandmath.files.wordpress.com/2007/09/statisticswomen.pdf.

[23]For information about the different national organisations, see https://www.mathunion.org/cwm/organizations/country.

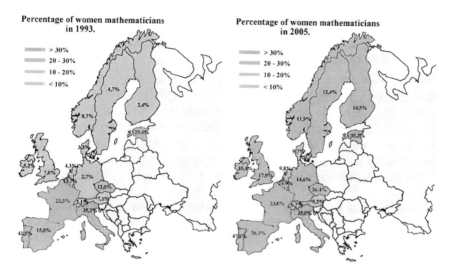

Fig. 5 Percentage of women mathematicians in Europe in 1993 and 2005

In 2015, with a remit to promote international contacts among all of these organisations, the IMU Committee for Women in Mathematics (CWM) was founded. The CWM website provides information about all the national organisations and much more besides, including links to articles about the recent history of women in mathematics.[24]

7 The Last Word Goes to Ingrid Daubechies

In 2010, Ingrid Daubechies was elected as the first woman president of the International Mathematical Union, her term of office running from 2011–2014. On 29 July 2014, just prior to the ICM in Seoul, Daubechies was interviewed for The World Academy of Sciences. Among the questions she was asked was[25]:

> There's a common assumption that women are less good than men at mathematics. What could be the reason for this, assuming it is true?

Here is Daubechies' response (her emphasis):

> I disagree with this view—*completely*. There is a highly variable percentage of women in academia and in departments of mathematics across Europe. Differences are so enormous that it becomes obvious that it has something to do with cultural habits, which differ from one nation to another, and not with intelligence. I have a very cynical colleague who says

---

[24]https://www.mathunion.org/cwm.

[25]https://twas.org/article/maths-also-women.

that the number of women mathematicians who are in academia is inversely proportional to some average of the amount of money and prestige that the job can grant: If there is little money and no prestige, there you'll find more women. I agree: These aspects seem to play a much larger role than being smart.

Daubechies draws attention to an important issue. She points out, and the historical examples support it, that even in a highly theoretical field such as mathematics, with often flexible working environments and relatively clear criteria for quality of work, social and political conditions impede the equality of women. But this very fact gives cause for hope too: these are conditions that can be changed and are being so, albeit not as fast as they might be.

We are not condemned to the repetition and perpetuation of past mistakes.

Acknowledgements I thank Deborah Kent, Caroline Series, and Reinhard Siegmund-Schultze for their valuable comments and suggestions which helped me produce an improved final version of this article.

References

1. Blum, L. (1991). 'A Brief History of the Association for Women in Mathematics: the Presidents' Perspectives' *Notices of the American Mathematical Society* **38**, 738–774.
2. Findlen, F. (2011). 'Calculations of faith: mathematics, philosophy and sanctity in 18th-century Italy (new work on Maria Gaetana Agnesi)' *Historia Mathematica* **38**, 248–291.
3. Montucla, J.-E. (1799–1802) *Histoire des Mathématiques*, Volume 2, p. 171.
4. Lovelace, A.A. (1843) 'Notes by A.A.L. [August Ada Lovelace]' *Taylor's Scientific Memoirs* **III**, 666–731.
5. Hollings, C., Martin, U., Rice, A. (2017). 'The Lovelace-De Morgan Correspondence: A critical reappraisal' *Historia Mathematica* **44**, 202–231.
6. Somerville, M. (1873). *Personal Recollections from Early Life to Old Age of Mary Somerville*, John Murray, 1873, p. 172.
7. Somerville, M. (1826). 'On the magnetizing power of the more refrangible solar rays' *Philosophical Transactions of the Royal Society of London* **116**, 132–139.
8. Mason, J. (1992). 'The Admission of the First Women to the Royal Society of London' *Notes and Records of the Royal Society of London* **46**, 279–300.
9. Koblitz, A.H. (1983)/ *A Convergence of Lives. Sofia Kovalevskaia: Scientist, Writer, Revolutionary*, Birkhäuser, pp. 215–217.
10. Kovalevsakya, S. (1978) 'My Recollections of George Eliot' in R. Chapman, E. Gottlieb 'A Russian View of George Eliot' *Nineteenth Century Fiction* **33**, p. 359.
11. Kovalevskya, S. (1895). *Sónya Kovalévsky. Her Recollections of Childhood* (tr. I. F. Hapgood), The Century Company (New York), p. 219.
12. Kovalevskaya, S. (1978) *A Russian Childhood* (tr. B. Stillman), Springer-Verlag, p.ix.
13. Kaufholz-Soldat, E. (2017). '[...] the first handsome mathematical lady I've ever seen!' On the role of beauty in portrayals of Sofia Kovalevskaya', *BSHM Bulletin* **32**, 198–213.
14. Loria, G. (1903)'Les Femmes Mathématiciennes' *Revue Scientifique* **20**, 385–392.
15. Cooke, R. (1984). *The Mathematics of Sonya Kovalevskaya*, Springer-Verlag, pp. 175–176.
16. Cooke, R. (1984). *The Mathematics of Sonya Kovalevskaya*, Springer-Verlag.
17. Laubenbacher, R., Pengelley, D. (2010). "Voici ce que j'ai trouvé:" Sophie Germain's grand plan to prove Fermat's Last Theorem, *Historia Mathematica* **37**, 641–692, p. 644.

18. Laubenbacher, R., Pengelley, D. (2010). "Voici ce que j'ai trouvé:" Sophie Germain's grand plan to prove Fermat's Last Theorem, *Historia Mathematica* **37**, 641–692, p. 690.
19. Anon. (1880). *The Spectator*, 7 February 1880.
20. Anon. (1890). Miss Fawcett's Honor: the sort of girl this Lady Senior Wrangler is, *The New York Times*, 8 June 1890, p.5.
21. Cartwright, M.L. (1944). Grace Chisholm Young, *Journal of the London Mathematical Society* **19**, 185–192, pp.185–187.
22. Grattan-Guinness, I. (1972). A mathematical union: William Henry and Grace Chisholm Young *Annals of Science* **29**, 105–185, p. 141.
23. Rothman, P. (1996) 'Grace Chisholm Young and the division of laurels' *Notes and Records of the Royal Society* **50**, 89–100.
24. Grattan-Guinness, I. (1975) 'Mathematical bibliography for W.H. and G.C. Young' *Historia Mathematica* **2**, 43–58, p. 43.
25. Mihaljević-Brandt, H., Santamaría, L., Tullney, M. (2016) 'The Effect of Gender in the Publication Patterns in Mathematics'. PLoS ONE 11(10): e0165367. https://doi.org/10.1371/journal.pone.0165367.
26. Kant, I. (1960). *Observations on the Feeling of the Beautiful and Sublime*, tr. J.T. Goldthwait, University of California Press, p.79. Originally published in German in 1764.
27. Tobies, R. (in press) 'Internationality. Women in Felix Klein's Course at the University of Göttingen 1893–1920'.
28. Siegmund-Schultze, R. (2018) 'Emmy Noether: "The Experiment to Promote a Woman to a Full Professorship" ', *Newsletter of the London Mathematical Society*, May 2018, 24–29, p. 27.
29. Green, J., LaDuke, J. (2008). *The Pioneering Women in American Mathematics. The Pre-1940s PhDs*, American Mathematical Society.
30. Greenwald, S.J. et al. (2015). 'The Association for Women Mathematicians: How and Why It Was Founded and Why It's Still Needed in the 21st Century', *The Mathematical Intelligencer* **37**, 11–21, p.13.
31. Bayer-Fluckiger, E. (1994). 'Women and Mathematics' in *First European Congress of Mathematics: Paris, July, 6–10, 1992. Round Tables*, Birkhäuser, pp. 37–74.

Initiatives of the International Union of Pure and Applied Physics to Reduce the Gender Gap in Physics

Silvina Ponce Dawson

Abstract The International Union of Pure and Applied Physics (IUPAP) was established in 1922 with 13 member countries. Nowadays, the physics communities of 54 countries are represented at the union. The IUPAP is governed by its General Assembly (GA) which meets triennially. Its top executive body is the Council which supervises the activities of commissions that are organized around physics sub-fields and of temporary working groups that are meant to solve specific problems. The 1999 IUPAP GA approved the creation of the Working Group on Women in Physics to survey the situation of women physicist and to suggest measures to improve it. The Working Group has performed a variety of actions to fulfill its mandate. Based on its recommendations, the IUPAP has taken a series of actions to reduce the gender gap and increase diversity and inclusion in physics. In this paper I briefly describe some of these actions.

1 Introduction

The International Union of Pure and Applied Physics (IUPAP) was established in 1922 with 13 member countries. Nowadays, the physics communities of 54 countries are represented at the union. The aims of the IUPAP are: to stimulate and promote international cooperation in physics; to sponsor suitable international meetings and to assist organizing committees; to foster the preparation and the publication of abstracts of papers and tables of physical constants; to promote international agreements on other use of symbols, units, nomenclature and standards; to foster free circulation of scientists; to encourage research and education. The Union is governed by its General Assembly (GA) that meets triennially. Its executive body is the Council which supervises the activities of twenty specialized International

S. P. Dawson (✉)
Departamento de Física, FCEN-UBA, Ciudad Universitaria, Buenos Aires, Argentina

IFIBA, CONICET-UBA, Ciudad Universitaria, Buenos Aires, Argentina
e-mail: silvina@df.uba.ar

Commissions and four Affiliated Commissions, each of which are organized around physics sub-fields. Besides the commissions, there are (temporary) working groups that focus on developing new research fields and activities that would be difficult to resource through traditional funding programs.

In 1999, the IUPAP GA approved the creation of the Working Group on Women in Physics with the following mandate: to survey the situation for women in physics in IUPAP member countries; to analyze and report the data collected along with suggestions on how to improve the situation; to suggest ways that women can become more involved in IUPAP; to report all findings at the General Assembly in 2002. Since then, the Working Group has engaged in a variety of actions to fullfill its mandate. In spite of this, the gender gap in physics still exists in most countries. For this reason, all IUPAP GAs since 1999 have approved the continuing existence of the Working Group on Women in Physics and the last one extended its mandate for six years, until 2023.

One of the first actions of the Working Group was to organize the constitution of working groups in as many countries as possible, with the task of collecting local information on the situation of women physicists. It also subcontracted the American Institute of Physics to perform a survey on the situation of women physicists. The questionnaire for this survey, in English, was disseminated via e-mail and was responded by about 1000 women physicists from 55 countries. All this information was presented at the First International Conference on Women in Physics (ICWIP) that took place at the UNESCO premises in Paris, France, in 2002 with over 300 participants from 65 countries. As we describe in what follows, this was the first in a series of International Conferences and of other activities organized by the Working Group to make more visible the contributions of women physicists, to help the professional development of female physicists in the early stages of their careers and to attract more girls to the discipline.

Besides the specific actions that the Working Group carried on to analyze and reduce the gender gap in the discipline, its most important contribution has been to put the issue of women in physics on the agenda of physics communities across the world and to help the creation of an international network of women physicists that extends over many more countries than IUPAP members. We describe in what follows some of the activities of the Working Group and of the decisions that, based on the recommendations of the Group, the IUPAP has taken over the years to increase diversity and inclusion in the practice of physics.

2 Activities of the IUPAP Working Group on Women in Physics

We describe in this Section the main activities that the Working Group (WG) on Women in Physics (WiP) has recently been engaged in, namely, to organize an International Conference on Women in Physics, ICWIP, every other three years; to elaborate resolutions and recommendations to be presented at the IUPAP GA; to give out travel grants to women physicists in the early stages of their careers and physics graduate students from developing countries; to provide useful information and to take the necessary actions to continually survey the situation of women physicists across the globe. The latter led the group to liaise with representatives from other international scientific unions and elaborate the project entitled "A Global Approach to the Gender Gap in Science: How to Measure It? How to Reduce It?" that received one of the 3 grants of the International Science Council (ICS, former ICSU) in 2016. This on-going project had a Global Survey of Scientists open through the end of 2018 which collected over 35,000 responses.

2.1 International Conferences on Women in Physics

The International Conferences on Women in Physics, ICWIP, constitute the main forum for discussing ways to improve the situation and increase the number of women physicists, for exchanging ideas on science and gender-related issues and for learning from regional differences. They provide the ideal platform to establish mentoring schemes and disciplinary and regional networks. Attendance is by country teams, with limits to the number of participants per country so that all countries are more or less equally represented. Travel grants are awarded to representatives from developing countries to guarantee ample participation. The attendance of graduate students and of at least one man per country team is encouraged.

Organizing each ICWIP poses a great challenge with respect to funding. Although the IUPAP sponsors the conference, the sponsorship is not enough to fully run it. The local organizers need to raise local funds to cover the costs of the organization. The WG, which acts as the International Organizing Committee, raises the money to cover the travel expenses of representatives from countries in need. Typically, about 70 grants are awarded. Grants cover both travel and lodging expenses of the participants.

Each ICWIP is organized every other three years in different parts of the world. The 1st ICWIP that took place in Paris, France, in 2002 was followed by conferences in Rio de Janeiro, Brazil, in 2005; Seoul, Republic of Korea, in 2008; Stellenbosch, South Africa, in 2011; Waterloo, Canada, in 2014 and Birmingham, UK, in 2017. The forthcoming ICWIP will be held in Melbourne, Australia, in 2020.

Lately, an ICWIP lasts for about 3.5 days and includes six main types of activities. *Plenary talks* have the purpose of making visible the contributions of women physicists. In almost all cases they are given by women physicists who describe their research intermixed with their personal life history. *Country poster sessions* are the event at which each country team must present the description of their local situation including a comparison with what had been presented in previous years. The *scientific poster session* is meant to facilitate the scientific exchange among attendants of the conference. Participation in this session is voluntary. In recent years, both poster sessions have been complemented with advertisement sessions composed of 3 min presentations of all posters. Five *workshops or break-out sessions* are run in parallel to address gender-related topics. Some of them are intended to provide tools for the professional development of physicists in the early stages of their careers. Each workshop typically has 3 sessions of 1.5 h each with a few short talks and ample time for discussion. The last session is devoted to elaborate recommendations to be discussed at the final *Conference Assembly*. During this final assembly, all participants discuss and draft the recommendations and resolutions to be presented at the IUPAP GA. Each ICWIP has at least one *outreach activity* for the general public or for school kids.

The themes discussed at the parallel workhops have changed a little since the first ICWIP. Among others, we have had sessions on: Attracting Girls into Physics; Launching a Successful Physics Career; Getting Women into the Physics Leadership Structure Nationally and Internationally; Improving the Institutional Climate for Women in Physics; Learning from Regional Differences; Balancing Family and Career; Professional Development and Leadership; Leaving and entering a career in physics; Gender studies and Intersectionality; Scientific networking in disciplines Physics and Science Education; Cultural perception and bias; Science Practice and Ethics.

All ICWIPs have had proceedings containing country papers (a brief written account of the country posters), papers associated to the workhops and other related material such as recommendations and resolutions. All the proceedings have been published as an issue in the AIP Conference Proceedings series.

To give a feeling of what the conferences are like, we give here a brief account of the last ICWIP. It was held in 2017 in Birmingham, UK, and had 212 participants representing 48 countries. 57 participants were awarded travel grants, although only 49 were able to attend (most of them had difficulties with visas). Nobel Peace Laureate Malala Yousafzai spoke as a guest at the conference and interacted with delegates. There was a special art exhibit, "Finding Space", that featured photos of the delegates in a soundscape created from sounds recorded at the research labs of female physicists. Immediately after the closing of the conference, attendants could choose to participate in continual professional development workshops on how to engage girls with science qualifications and careers, on unconscious bias or on how to get gublished. Travel grant holders had the chance to spend an extra day visiting labs and theoretical groups of the University of Birmingham to foster collaborations with the researchers. Also for ICWIP17 we faced a recurrent problem of international conferences: the denial of visas. Seven visas were denied or delayed preventing participation in ICWIP17.

2.2 Travel Grants for Young Female Physicists and Physics Students

The years with no ICWIPs, the funds that the WG receives are mostly spent on travel grants for female physics students and physicists in early stages of their careers. The grants are given to women from developing countries that are willing to attend a conference, school or workshop outside their home institution. The WG makes the selection of the awardees. In 2012, 83 applications were received and 22 grants were awarded to women from Argentina, Brazil, China, Ghana, Hungary, India, Iran, Kazakhstan, Kenya, Peru, Poland, Romania and Ukraine. In 2013, the number of applications was 64 and the number of grants 15; they were, respectively, 100 and 26 in 2015; 30 and 23 in 2016.

2.3 Other Activities of the Working Group

The WG has a webpage, currently hosted at wgwip.df.uba.ar, with useful information for the women in physics community. The page contains information on the WG (current and past); the list of country team members and team leaders with contact information; the resolutions and recommendations elaborated at the the various ICWIPs; the resolutions suggested by the WG that were approved by the IUPAP GAs; information on all ICWIPs and the Travel Grants awarded by the WG; links to useful resources; some news such as the calls for travel grant applications. Currently, there is not a communication group fully devoted to maintain the website.

The WG decided to establish an International Women in Physics Day. This resolution was approved by the IUPAP GA in 2014. Some discussion was then started about the date. The WG finally chose February 11th as the International Women in Physics Day coinciding with the International Day of Women and Girls in Science as established by the United Nations. In 2018, the WG encouraged the participation in the campaign launched by the Organization for Women in Science for the Developing World (OWSD) portraying stories of female physicists and physics students on social networks. The WG is planning to fully launch the International Women in Physics Day in 2019, the year that marks the 20th anniversary of the IUPAP General Assembly that decided on the group's creation. In preparation for this, the group is planning to open a competition for the design of a logo to identify the Day. A redesign of its web page adding new capabilities is also being planned.

During the 5th ICWIP that took place in Waterloo, Canada, in 2014, the decision was made to write a declaration of principles drawing inspiration from the Baltimore Charter of the American Astronomical Society. This gave birth to the *Waterloo Charter on Women in Physics* which is a declaration of principles endowed with a long list of recommendations to advance towards a more inclusive and diverse practice of physics. It is based on the rubrics of the Baltimore Charter and the

Pasadena Recommendations formulated by the American Astronomical Society in 1993 and 2003 respectively. It is also shaped and guided by the principles dictated by the JUNO project of the Institute of Physics in the UK. It embraces, as well, the statements on gender equity and inclusiveness in physics that have been issued previously by the IUPAP. The final draft is currently under the consideration of the IUPAP Executive Council to be presented at the next GA that will be held in 2020. This latest draft is available at the WG webpage.

2.4 Resolutions and Recommendations Presented by the Working Group on Women in Physics

Several resolutions and recommendations were generated, between 2002 and 2017, at the 6 ICWIPs. Recommendations were intended for individuals, physical societies, the WG, the IUPAP and other players of the scientific endeavor. In this section we quote some of them.

Scientific and Professional Societies should foster gender equity having a group examining policies, making available statistics on the participation of women, identifying leading women physicists and promoting them as role models

Funding Agencies should ensure that there is no gender bias in the broad based general grant funding process, and that women are included on review and decision making committees. Limits on age of eligibility or grant duration that seriously disadvantage applicants taking family leave should be reconsidered. Statistics should be made available giving by gender the proportion of successful applicants.

All institutions should note that family oriented practices such as flexible work schedules, opportunities for dual career families, and child care facilities increase the opportunities for women in science and technology.

The IUPAP should require the organizers of the conferences it sponsors to improve the inclusion and encouragement of women, and request its member societies and other scientific unions to do the same.

The IUPAP should actively encourage the organizers of the conferences it sponsors to provide, associated with the program: (a) professional development workshops for attendees and (b) outreach activities aimed at the public and to engage both girls and boys from an early age in the excitement of physics.

The IUPAP should encourage its commissions and member countries to have a clear and transparent list of criteria for awards ensuring that women are nominated for prizes and that there are women on the selection committees for prizes and awards.

2.5 The Global Survey of Physicists

The Global Survey of Physicists was open for responses between 2009 and 2010. It was available in 8 languages. The survey was implemented and analyzed by the Statistical Research Center of the American Institute of Physics who had done the previous (smaller and women only) surveys in 2001–2002. The questionnaire was developed by the AIP in close collaboration with the WG and some country team leaders.

The survey collected almost 15,000 responses from about 130 countries, 75% of them, highly developed. 21% of the responses coming from developed countries were from female respondents, this percentage was 27% in the case of less developed countries. Less than 30% of the responses came from (mostly graduate) students.

The survey confirmed some of the results that had been obtained in 2001–2002. For example, it showed that early educational experiences had been important for both men and women in choosing physics, especially in highly developed countries. It also showed differences between genders and between more or less developed countries. Regarding their personal lives, male physicists were more likely than women to have spouses that earn less than they do, who don't work or who do most of the housekeeping. The difference was most noticeable in the case of highly developed countries. For example, while only 7% of the female respondents from developed countries and 5% from developing ones said that their spouses were not employed, the ratio was 33% of male respondents from developed countries and 23% from developing ones. There were also noticeable differences with respect to having kids. While about 40% of the female respondents from developing countries had had their children after university and before their doctorate only 25% of the men from developing countries and 15% of the women and 10% of the men from developed ones had had their kids at that stage in their careers. Sixty percent of the female and 65% of the male respondents from developed countries had their kids after their doctorate. These data highlighted differences in family organization and family related decisions that were likely to interfere more with the advancement of women's than of men's careers.

Some of the responses also reflected differences in attitudes. While about 50% of the female respondents from developing countries said they felt mostly comfortable raising concerns with their boss 59% of male respondents from developing countries and 61% and 70%, respectively, of female and male respondents from developed ones gave the same answer.

Finally, the information that the respondents provided about their careers as scientists showed that women had a harder time than men finding opportunities to advise graduate students, to serve as journal editors or on influential committees, to have an international work experience and to receive invitations to speak.

2.6   Measures Approved by the IUPAP to Increase Inclusiveness and Diversity in Physics

Besides the creation and continuous renewal of the WG on WiP, the IUPAP has taken other measures to advance the agenda of women in physics. In 2011, it created the position of Gender Champion and decided that one Vice-President at Large would be assigned to occupy this position. The person occupying this position is elected at the GA. The position was proposed to strengthen connections between the IUPAP commissions and the Working Group on Women in Physics. Since that time, every three years a Vice President at Large has been appointed to serve in that position mainly to assist in tracking the representation of women in all IUPAP activities. Based on the collected data, a set of rules has been established to guarantee that women are represented as organizers, speakers and attendees of IUPAP sponsored and supported conferences and that conference participants receive information on inclusiveness in physics. An anti-harassment policy has also been established for such conferences.

The 29th GA established, as a recommendation for all affiliated national institutions, that a 20% target (of female participation in conferences) be achieved. It has also been defined that meetings with female participation of less than 10% are not accepted. Conference organizers will have a deadline of a few weeks to make the necessary corrections if this 10% is not fulfilled. An analysis of the conferences sponsored by the IUPAP in 2017 and 2018 showed an average of 19% female participants and 17% of female invited speakers, not too far away from the desired (minimum) target of 20%.

The 29th GA also passed a resolution that addressed the "need to encourage IUPAP-sponsored conferences to have a session for all participants on Diversity and Inclusion in Physics". The Executive Council has discussed examples of possible activities. They include: a plenary session or talk, a brief presentation followed by an exhibition that remains open a significant amount of time during the conference, a survey on the issue to be responded by conference participants, etc.

IUPAP requires that supported conferences publish on their websites and in all publications related to the Conference a specific statement on harassment. Among other things, the statement says: "The conference organisers will name an advisor who will consult with those who have suffered from harassment and who will suggest ways of redressing their problems, and an advisor who will counsel those accused of harassment."

2.7   Partnering Up with Other Scientific Unions

The problems that women physicists face are shared by women in most other careers in STEM. The need to have consistent data on the situation of gender in physics and other disciplines and to elaborate policy proposals supported by data was the initial seed for the project entitled "A Global Approach to the Gender Gap in Mathematical and Natural Sciences: how to measure it, how to reduce it?". This collaborative project among 8 scientific unions with the leadership of the International Mathematical Union and the International Union of Pure and Applied Chemistry has three tasks. Task 1 involves performing the global survey of scientists that we referred to before. Task 2 consists of performing a study on publication patterns in the disciplines of the project. Task 3 consists of creating an online database with lists of good practices for girls, women, parents and organizations at various levels, including those involved in guiding young women into careers. IUPAP representatives are in the coordination group of Task 1. The IUPAP in general and its WG and Gender Champion, in particular, are fully committed to work on this project and contribute to achieve its goals of providing evidence on which to orient future actions to help reduce the gender gap in science, collaborate with social scientists to determine differences and commonalities across regions, cultures and disciplines cross-referencing the data with other available indicators about the countries, to provide easy access to materials to encourage young women to work in the fields of STEM and to recommend practical policies and actions to reduce the gender gap in science.

3 Final Words

Women have made significant and creative contributions in physics. These contributions, however, most often have gone unnoticed. This is reflected, for example, in that 2018 has seen the first female Nobel Prize winner in 55 years. The percentage of female physicists, on the other hand, remains low in many countries. It is increasingly clear that scientific careers are strongly affected by social and cultural factors, and are not determined solely by merit. The search for excellence that unites all scientists can be maintained and enhanced by increasing the diversity of its practitioners. The attainment of such diversity requires that criteria for judging excellence be free of cultural perceptions and bias.

Creating its Working Group on Women in Physics the IUPAP recognized that something had to be done to improve the situation and increase the number of female physicists. The existence of the Working Group brought the issue upfront and made the physics community aware that there was a problem that called for specific actions The actions of the WG led to the creation of a large network of women physicists and inspired the organization of several activities all over the world. In spite of all of this, much more still needs to be done. The IUPAP has recognized it through its deep involvement with the Gender Gap in Science Project and by having approved the WG existence for 2 more periods (six years).

The activities of the IUPAP WG on WiP are not only aimed at improving the situation of women physicists. On one hand, special attention is paid to the situation of developing countries and measures are considered to help them have a thriving physics community. On the other hand, improvements in the workplace environment, as those suggested by the WG, are beneficial for everbody, not only women. Increasing the number of people of all genders that study or work in physics has been a permanent concern of the group. That is why the WG has continually recommended the organization of outreach activities and has been directly engaged in their organization at ICWIPs.

Reducing the gender gap and increasing diversity and inclusiveness in science requires a cultural change. We all have to be part of a worldwide concerted effort for the necessary changes to enter into effect.

Acknowledgements I thank Marie-Francoise Roy for the invitation to be part of the panel "The Gender Gap in Mathematical and Natural Sciences from a Historical Perspective" which took place during that International Congress of Mathematicians (ICM), Rio de Janeiro, Brazil, in August, 2018. My participation on this panel was possible thanks to the financial support of the project entitled "A Global Approach to the Gender Gap in Science" and the CWM. Further support from UBA (UBACyT 20020170100482BA) and ANPCyT (PICT 2015–3824) is also acknowledged.

Printed in the United States
By Bookmasters